Researching and Writing in the Sciences and Technology

Christine A. Hult
Utah State University

✦ ✦ ✦

Allyn and Bacon
Boston ✦ London ✦ Toronto ✦ Sydney ✦ Tokyo ✦ Singapore

Executive Editor: Joseph Opiela
Editorial Assistant: Susannah Davidson
Executive Marketing Manager: Lisa Kimball
Editorial-Production Administrator: Catherine Hetmansky
Editorial-Production Service: Ruttle, Shaw & Wetherill, Inc.
Composition Buyer: Linda Cox
Manufacturing Buyer: Aloka Rathnam
Cover Administrator: Suzanne Harbison

Copyright © 1996 by Allyn & Bacon
A Simon & Schuster Company
Needham Heights, MA 02194

All rights reserved. No part of the material protected by this copyright notice may be reproduced or utilized in any form or by any means, electronic or mechanical, including photocopying, recording, or by any information storage and retrieval system, without the written permission from the copyright holder.

Library of Congress Cataloging-in-Publication Data
Hult, Christine A.
    Researching and writing in the sciences and technology/Christine A. Hult.
    p.   cm.
    Includes bibliographical references and index.
    ISBN 0-205-16840-X
    1. Research—Methodology. 2. Technical writing. I. Title.
Q180.55.M4H85   1995
507′.2–dc20   95-21280
                                                                                  CIP

Printed in the United States of America
10 9 8 7 6 5 4 3 2 1   00 99 98 97 96 95

Contents
✦ ✦ ✦

PREFACE vii

✦ CHAPTER 1   College Research   1

The General Process of Research   1
- *Preparation*   2
- *Incubation*   2
- *Illumination*   3
- *Verification*   3

A Physician Uses the Research Process   4
The Importance of Observation in the Social Sciences   6
The Importance of Formulating and Testing Hypotheses   7
Critical Scientific Research   8
The Importance of Replicability and Scientific Debate   9

✦ CHAPTER 2   Library Resources   11

The Library Reference Area   11
Online Computer Catalogs   13
- *How to Search the Online Catalog*   13
- *Internet*   14
- *Subject Headings*   16
- *Recording Bibliographic Information*   19

Microforms   20
Locating Articles in Serials: Popular Periodicals   20
- *Newspapers*   20

Locating Articles in Serials: Professional Journals   21
Locating Government Documents   23
Other Technologies   24
- *Online Database Services*   24
- *CD-ROM Searching*   24

Accessing Information from Other Libraries   26
Discipline-Specific Resources for Science and Technology   26

iii

iv   Table of Contents

✦ CHAPTER 3   Library Research Methods   33

Preparation and Incubation   33
    Finding a Topic   33
    Gathering Research Materials   34
    The Working Bibliography   35
    Developing a Search Strategy   37
    Outlining a Time Frame   39

Locating Sources   41

Working with Sources   41
    Reading for Meaning   41
    Active Reading and Notetaking   43
    Paraphrasing Appropriately and Avoiding Plagiarism   43
    How to Paraphrase Appropriately   46
    Making Section-by-Section Summaries   48
    Reviewing   49
    Perceiving the Author's Organizational Plan   49

Illumination and Verification   54
    Evaluation   54
    Writing from Sources   55
    Summarizing   56
    Synthesizing   57
    Critiquing   59

✦ CHAPTER 4   Primary Research Methods: Writing a Research Report   63

Primary Research in the Sciences   63
    Lab Experiments and Reports   63
    Field Observations and Reports   64

Sample Scientific Report: Biology   67

Research Reports in Technology and Engineering   81

Sample Research Report: Engineering and Technology Format   82

✦ CHAPTER 5   Planning, Writing, and Revising Your Research Paper   101

Rhetorical Situation   101
    Purpose   101
    Persona   102
    Audience   102

Subject Matter   103
Appropriate Language or Tone   103

Organization   104

Revising Your Research Paper   104
    Revising for Structure and Style   105
    Improving Paragraphs   106
    Improving Sentences   107
    Improving Words   107
    Editing for Grammar, Punctuation, and Spelling   110
    Mechanical Errors   111

Rewriting Your Paper Using Word Processing   115
    Revising with Word Processing   115
    Editing with Word Processing   115
    Grammar-Checking Software   116
    Bibliography and Footnote Software   116
    Incorporating Reference Materials   117
    Incorporating Direct Quotes   117
    Formatting and Printing Using Computers   119
    Proofreading   119

Considering Formal Details   120
    Spacing   121
    Margins   121
    Title   121
    Numbering   122
    Indentation   122
    The Abstract   122
    The Endnote Page   123
    The References Page   123
    The Annotated Bibliography   123
    The Appendix   124
    Using Graphics   124

✦ CHAPTER 6   Secondary Research Methods: Writing a Review Paper   127

A Guide to the Scientific Review Paper   128
    Preparation   128
    Developing a Search Strategy   128
    Focusing Your Search   130

Evaluation   135

Organizing and Writing the Scientific Review Paper   136

> *Arranging the Materials* 137
> *Writing the First Draft: Verification* 138

Documentation in Science and Technology 139
> *Internal Citation* 139
> *The Reference List* 142

Model References: Natural and Physical Science (CBE) 143
Exercises and Research Project 149
Sample Review Paper: Science Format (CBE) 150

INDEX 162

Preface
✦ ✦ ✦

Researching and Writing in the Sciences and Technology is an interdisciplinary research text that introduces you to research processes used in the sciences (such as Biology and Geography) and technology (such as Engineering and Industrial Technology). By reading this book you will gain experience in posing and solving problems common to an academic discipline, learning both primary research strategies and library research strategies. A comprehensive list of library resources is included to provide you with access to the important tools used by researchers.

Also included, as examples, are model student reports and research papers from the sciences and technology to show you how your peers have solved research problems similar to your own. The exercises are designed to guide you through research processes and to teach the important supporting skills of summarizing, synthesizing, and critiquing source materials. Complete listings of citations show you how to document your sources within each discipline. In addition, this textbook stresses principles of research presentation and documentation common to the sciences and technology.

Many of the research guides now on the market fail to provide the broad introduction to research that students need. Traditional research texts are often focused very narrowly on the library "term paper." They discuss formal considerations at length, from the form of notecards and bibliography cards to the form of a completed term paper. But they do not explore the entire research process, which is an integral part of any successful research project.

In contrast, *Researching and Writing in the Sciences and Technology* is a research text that explains and fosters intellectual inquiry, compares research in disciplines allied with the sciences and technology, and provides you with logical practice in research methodology.

Researching and Writing in the Sciences and Technology is divided into chapters of general information about research methods and resources in the academic disciplines and specific guidance in writing research papers for courses in the sciences and technology. You are first introduced to college research in general, and then to library resources (both general and discipline specific), library research methods, and primary research methods. Next, you will receive explicit instruction in the planning, writing, and presentation of research papers in any field.

Exercises throughout the book are designed to help you conduct your own research projects in a systematic and organized fashion.

Researching and Writing in the Sciences and Technology is both a comprehensive guide to research processes and an easy-to-use, complete reference tool designed to be used throughout students' academic career and into their professional life. As with all research, my own work on this book has been a challenging process of discovery. I am grateful to the many researchers and theorists in the field of composition and rhetoric on whose work this book is built. Although I have cited in chapter notes only those authors whose ideas directly contributed to my own, many others contributed their ideas indirectly through journals, conferences, and textbooks. Teachers I have studied under and worked with, students who have patiently tried my ideas, and friends and family who have supported me along the way have all helped in the genesis of this book.

Finally, I am grateful to the editors and production team at Allyn & Bacon for their personal, professional attention, and to the reviewers of this edition of my manuscript.

1

College Research

In the Western world, the sciences hold an authoritative position and are a dominant force in our lives. The sciences have been enormously successful at formulating and testing theories related to phenomena in the natural and physical world. These theories have been used to solve physical and biological problems in medicine, industry, and agriculture. Generally, the sciences have been divided into life sciences (such as zoology and botany), physical sciences (such as physics and chemistry), and earth sciences (such as geology and geography). Scientific insights and methods have also been carried over to fields of applied science and technology, such as computer science and engineering.

Research, broadly defined, is systematic inquiry designed to further our knowledge and understanding of a subject. Using this definition, nearly everything you do in college is "research." You seek to discover information about people, objects, and nature; to revise the information you discover in light of new information that comes to your attention; and to interpret your experience and communicate that interpretation to others. This is how learning proceeds both for all of us as individuals and for human beings together as we search for knowledge and understanding of our world.

THE GENERAL PROCESS OF RESEARCH

Many accounts of the process of research have been written by scientists, artists, and philosophers. Researchers generally agree about the outlines of the process regardless of the discipline. The process often begins with a troubled feeling about something observed or experienced followed by a conscious probing for a solution to the problem, a time of subconscious activity, an intuition about the solution, and finally a systematic testing to verify the solution. This process may be described in a number of ways but may generally be divided into four stages: preparation, incubation, illumination, and verification.[1] These stages are discussed in order, but during a research project, each stage does not necessarily neatly follow the next. Quite probably the researcher moves back and forth freely among the stages and may even skip a stage for a time, but generally these four stages are present in most research projects.

Preparation

The preparation stage of the research process involves the first awareness in researchers that a problem or a question exists that needs systematic inquiry. Researchers formulate the problem and begin to explore it. As they attempt to articulate the exact dimensions and parameters of a particular problem, they use language or symbols of their discipline that can be more easily manipulated than unarticulated thoughts or the data itself. By stating the problem in a number of ways, looking at it from various angles, trying to define its distinctive characteristics, and attempting possible solutions, researchers come to define for themselves the subtleties of the problem. Preparation is generally systematic, but it may also include the researchers' prior experiences and the intuitions they have developed over time.

Incubation

The incubation stage usually follows the preparation stage and includes a period of intense subconscious activity that is hard to describe or define. Because it is so indistinct, people tend to discount it as unimportant, but the experience of many researchers shows that it is crucial to allow an idea to brew and simmer in the subconscious if a creative solution is to be reached. Perhaps you have had the experience of trying and failing to recall the name of a book you recently read. You tell your conversation partner, "Go ahead. It'll come to me." A few minutes later, when you are not consciously trying to recall the name but have gone on to other matters in your conversation, you announce, *"Silent Spring,"* out of the blue. This is an example of the way your subconscious mind continues to work on a problem while your conscious mind has gone on to another activity.

We need to allow ourselves sufficient time for incubating. If you have ever watched a chicken egg in an incubator, you will have a sense of how this works. The egg rests in warmth and quiet; you see no action whatsoever, but you know that beneath that shell tremendous activity is going on. The first peck on the shell from the chick about to hatch comes as a surprise. This is analogous to the next stage of the process, illumination. You cannot really control the incubation of a problem, but you can prepare adequately and then give yourself enough time for subconscious activity.

Illumination

In the illumination stage, as with the hatching chick, there is an imaginative breakthrough. The idea begins to surface out of its concealing shell, perhaps a little at a time. Or the researcher leaps to a hypothesis, a possible solution to the problem, that seems intuitively to fit. Isaac Newton discovered the law of universal gravitation as he watched an apple fall, and Archimedes deduced his principle of the displacement of water while in the bathtub. The illumination of a hypothesis can come suddenly or gradually, after laborious effort or after an ordinary event that triggers the researcher's thinking along new lines. We must remember, though, that the hypothesis comes only after the researcher has investigated the problem thoroughly. The egg must be prepared, fertilized, warmed, and cared for. The solution to a complex problem will come only after much conscious study and preparation in conjunction with subconscious intuition. Sometimes the solution will not be so much a breakthrough as it will be a clearer understanding of the problem itself.

Verification

Once a researcher has arrived at a hypothesis, he or she must systematically test it to discern whether it adequately accounts for all occurrences. Sometimes, this testing requirement necessitates a formal laboratory research experiment; in other cases, only an informal check against the researcher's own experience is necessary. In the sciences, the verification stage tends to be highly rigorous, involved, and lengthy. One should also be prepared at the verification stage to discover that the original hypothesis will not work. Although we are often reluctant to make mistakes, without a willingness to err we would never be led to make an original contribution. Research often progresses as a series of increasingly intelligent mistakes through which the researcher ultimately is led to a reasonable and workable solution. Sometimes hypothesis testing goes on for years and, for a particularly promising hypothesis, is performed by the research community in general. To be judged as sound, or verified, such a hypothesis must survive the critical scrutiny of the whole research community.

A PHYSICIAN USES THE RESEARCH PROCESS

In the following essay, Charles Nicolle, a physician and scientist of the early twentieth century, describes the research process he used to discover the mechanism that transmitted the disease typhus.[2] As you read the essay, pay particular attention to the research process Nicolle outlines.

The Mechanism of the Transmission of Typhus
Charles Nicolle

It is in this way that the mode of transmission of exanthematic typhus was revealed to me. Take all those who for many years frequented the Moslem hospital of Tunis, I could daily observe typhus patients bedded next to patients suffering from the most diverse complaints. Like those before me, I was the daily and unhappy witness of the strange fact that this lack of segregation, although inexcusable in the case of so contagious a disease, was nevertheless not followed by infection. Those next to the bed of a typhus patient did not contract the disease, while, almost daily, during epidemic outbreaks, I would diagnose contagion in the *douars* (the Arab quarters of the town), and amongst hospital staff dealing with the reception of patients. Doctors and nurses became contaminated in the country in Tunis, but never in the hospital wards. One day, just like any other, immersed no doubt in the puzzle of the process of contagion in typhus, in any case not thinking of it consciously (of this I am quite sure), I entered the doors of the hospital, when a body at the bottom of the passage arrested my attention.

It was a customary spectacle to see poor natives, suffering from typhus, delirious and febrile as they were, gain the landing and collapse on the last steps. As always I strode over the prostrate body. It was at this very moment that the light struck me. When, a moment later, I entered the hospital, I had solved the problem. I knew beyond all possible doubt that this was it. This prostrate body and the door in front of which he had fallen, had suddenly shown me the barrier by which typhus had been arrested. For it to have been arrested, and, contagious as it was in entire regions of the country and in Tunis, for it to have remained harmless once the patient had passed the Reception Office, the agent of infection must have been arrested at this point. Now, what passed through this point? The patient had already been stripped of his clothing and of his underwear; he had been shaved and washed. It was therefore something outside himself, something that he carried on himself, in his underwear, or on his skin, which caused the infection. This could be nothing but a louse. Indeed, it was a louse. The fact that I had ignored this point, that all those who had been observing typhus from the beginnings of history (for it belongs to the most ancient ages of humanity) had failed to notice the incontrovertible and immediately fruitful solution of the method of transmission, had suddenly been revealed to me. I feel somewhat embarrassed about thus putting myself into the picture. If I do so, nevertheless it is because I believe what happened to me is a very edifying and clear example, such as I have failed to find in the case of others. I developed my observation with less timidity. At the time it still had many shortcomings. These, too, appear instructive to me.

If this solution had come home to me with an intuition so sharp that it was almost foreign to me, or at least to my mind, my reason nevertheless told me that it required an experimental demonstration.

Typhus is too serious a disease for experiments on human subjects. Fortunately, however, I knew of the sensitivity of monkeys. Experiments were therefore possible. Had this not been the case I should have published my discoveries without delay, since it was of such immediate benefit to everybody. However, because I could support the discovery with a demonstration, I guarded my secret for some weeks even from those close to me, and made the necessary attempts to verify it. This work neither excited nor surprised me, and was brought to its conclusion within two months.

In the course of this very brief period I experienced what many other discoverers must undoubtedly have experienced also, viz. strange sentiments of the pointlessness of any demonstration, of complete detachment of the mind, and of wearisome boredom. The evidence was so strong, that it was impossible for me to take any interest in the experiments. Had it been of no concern to anybody but myself, I well believe that I should not have pursued this course. It was because of vanity and self-love that I continued. Other thoughts occupied me as well. I confess a failing. It did not arrest my research work. The latter, as I have recounted, led easily and without a single day's delay to the confirmation of the truth, which I had known ever since that revealing event, of which I have spoken. ✦

Nicolle's struggle to discover a solution to the problem of typhus transmission illustrates a research process. Nicolle worked on the problem of typhus transmission both consciously and subconsciously; his "Eureka" experience came unexpectedly and forcefully. Nicolle was awarded the Nobel Prize for medicine in 1928 for his discovery and for the experiments that conclusively confirmed that typhus was indeed transmitted by parasites.

✦ QUESTIONS FOR DISCUSSION

1. What stages of the research process outlined in this chapter are revealed in Nicolle's description?
2. What incongruity led Nicolle to research the question of typhus transmission?
3. What experience triggered the solution to the problem?

6 College Research

4. What procedures did Nicolle use to verify his hypothesis?
5. You may have noticed that Nicolle grew bored at points during his research. What kept him going?

✦ EXERCISES

1. The four research stages discussed in this chapter come from words we often associate with different contexts. For example, we use the term *preparation* in connection with preparing dinner or preparing for a test. First, briefly describe a situation (other than research) commonly associated with each term. Then list any similarities between the connotation of each word in your situation and the particular research stage.

   A. To prepare:
   B. To incubate:
   C. To illuminate:
   D. To verify:

2. Think of an activity with which you are familiar, such as a sport (football or tennis), a hobby (cooking or gardening), or an art (painting or dancing). In one paragraph, describe the process you use when participating in the activity and relate the process to the research stages discussed in this chapter (preparation, incubation, illumination, and verification).

THE IMPORTANCE OF OBSERVATION IN THE SCIENCES

As discussed earlier in this chapter, the motivation or impetus for much research is an observed event or experience that challenges our existing ideas and promotes inquiry. In the context of existing theories, such an event is incongruous and thus sparks in the researcher's mind a question or problem to be investigated. The researcher must be prepared to recognize the inconsistency and to see its importance. He or she must be familiar with current theories and concepts about the natural and physical world. In general, the aim of scientific work is to improve the relationship between our ideas (theories and concepts about the world) and our actual experiences (observations of the world).

An example of a scientist using educated observation is described by Rene Taton in his book *Reason and Chance in Scientific Discovery*.[3] In 1928, Sir Alexander Fleming, an English biologist, was studying mutation in some colonies of *Staphylococcus* bacteria. He noticed that one of his cultures had been contaminated by a microorganism from the air outside. But instead of neglecting this seemingly inconsequential event, Fleming went on to observe the contaminated plate in detail and noticed a surprising phenomenon: the colonies of bacteria that had been attacked by microscopic fungi had become transparent in a large region around the contamination. From this observation, Fleming hypothesized that the effect could be due to an antibacterial substance secreted by the foreign microorganism and then spread into the culture. Fortunately for us, Fleming decided to study the phenomenon at length to discover the properties of this secretion (which turned out to be a variety of the fungus *Penicillium,* from which we now make the antibiotic, penicillin) on cultures of *Staphylococcus* bacteria. Fleming designed experiments that tested his hypothesis concerning the effects of *Penicillium* on bacteria, and eight months later, he published his research findings in the *British Journal of Experimental Pathology.* Fleming's research is another example of the research process: first there was prepared observation, then a struggle with the problem and the formulation of a hypothesis, and finally the verification of that hypothesis using experiments.

THE IMPORTANCE OF FORMULATING AND TESTING HYPOTHESES

On the basis of a scientist's prior knowledge and preparation, he or she formulates a hypothesis to account for the observed phenomenon that presents a problem. Arriving at a hypothesis takes much effort on the part of the researcher. Brainstorming for possible hypotheses is an important component of research, because the researcher can creatively make conjectures based on prior experience. The researcher may have to test several possible hypotheses before deciding which one seems to account for the observed phenomenon. Fleming, for example, hypothesized that the clear circle he observed around the bacteria resulted from the contaminating microorganism in the culture. In the sciences, there are systematic ways of testing a hypothesis once it is formulated. Fleming used such a procedure to verify his hypothesis: he demonstrated through scientific experiments that *Penicillium* was effective against bacteria.

The following outline describes the systematic way, or *scientific method*, by which scientists customarily proceed:

1. The scientist formulates a question and develops a hypothesis that might shed light on the question posed.
2. On the basis of the hypothesis, the scientist predicts what should be observed under specified conditions and circumstances.

3. The scientist makes the necessary observations, generally using carefully designed, controlled experiments.
4. The scientist either accepts or rejects the hypothesis depending on whether or not the actual observations correspond with the predicted observations.

Using the scientific method, a researcher is able to integrate new data into existing theories about the natural and physical world. In step 2 above, the researcher draws on accepted scientific ideas and theories to predict an outcome for the experiment. Fleming, for example, predicted that his experiments would show the antibacterial action of *Penicillium* when used on certain bacteria, and he confirmed his hypothesis through careful laboratory experiments.

Neither Fleming nor the scientific community of his day recognized the profound implications of his research for the field of medicine. The antibiotic, *Penicillium*, was a difficult-to-handle, impure, and unstable substance, which at the time made it seem impractical for widespread application. Subsequent discoveries and refinements of antibiotics, however, proved that penicillin would revolutionize modern medicine.

Ordinarily, the individual researcher uses currently held scientific theories and ideas to incorporate new data into the mainstream of current scientific belief. Research advances are the cumulative result of researchers working on various problems in various parts of the world; a synthesis of existing data is used to create new ideas and theories. Taton notes that many of the immense scientific discoveries of the twentieth century have been the collective work of teams of specialists from various schools working with more and more nearly perfect technical resources.[4] Although the role of the individual researcher is important, he or she is but a cog in the wheel of the scientific community in general.

CRITICAL SCIENTIFIC RESEARCH

Much of the research conducted by scientists is an attempt to incorporate new data into existing theories, but another type of scientific research, critical scientific research, attempts to challenge currently held beliefs and theories in an effort to improve them. Critical scientific research investigates the adequacy, or the sufficiency, of theories about the natural and physical world. In this context, the question asked is, "How well do current theories actually explain the natural and physical world as we know and observe it?"

A particular field of science may operate for years under certain theoretical assumptions. For example, Newtonian physics, based on the theories of Isaac Newton, dominated the scientific world for some time, and thousands of scientists conducted regular experiments based on Newton's theories. But because physicists encountered numerous phenomena that were incompatible with Newton's laws of physics, new theories became necessary. The physicist Albert Einstein challenged the agreed-upon Newtonian physics by presenting an alternative system that accounted for more of the observed data. Most physicists have now adopted Einstein's more comprehensive theories or have gone on to develop and adopt new theories.

This process of challenging and replacing scientific theories is one way scientific fields advance their knowledge and understanding of the natural or physical world. In this way, scientific thought progresses both by regular scientific observation and experimentation, using widely accepted theories and beliefs, and by critical scientific research that challenges those widely held theories and suggests new ones.

THE IMPORTANCE OF REPLICABILITY AND SCIENTIFIC DEBATE

Scientists who have created and tested a hypothesis must then report their findings to other scientists, as Fleming did by publishing his experiments in the British journal. The goals of publishing one's findings include having other scientists accept the hypothesis as correct, communicating knowledge, and stimulating further research and discussion. A report of the research must necessarily include a careful, accurate description of the problem, the hypothesis, and the method (experimental design) used to test the hypothesis, in addition to the researcher's experimental findings and conclusions.

Other scientists then test the validity and reliability of the findings by attempting to repeat the experiment described by the researcher. A carefully designed and executed scientific experiment should be accurately described in writing so that other scientists using a similar experimental process can replicate it. The community of scientists as a whole then critiques the new research, deciding collectively whether or not it is good, sound research. To do this, other scientists will test the experiment's validity (Did it measure what the researcher said it would measure?), its reliability (Can it be repeated or replicated with similar results by other scientists?), and its importance (How does this experiment fit into a larger theoretical framework and what does it mean for our currently held assumptions and beliefs?). The forums of science—

the professional organizations and journals, universities, scientific societies, and research laboratories—combine to resolve scientific issues to the benefit of all scientists.

It is these forums of science that necessitate writing for scientists. The best research in the world will be of no consequence if the scientist is unable to communicate his or her results clearly to other scientists through writing. Two major forms of writing done by all scientists will be treated in this book: the *research report* (discussed in chapter 4) and the *review paper* (discussed in chapter 6).

NOTES

1. Adaptation of excerpt from "The Four Stages of Inquiry," in Richard E. Young, Alton L. Becker, and Kenneth L. Pike, *Rhetoric: Discovery and Change* (New York: Harcourt Brace Jovanovich, Inc., 1970). Reprinted by permission of the publisher. This four-step problem-solving process (preparation, incubation, illumination, verification) was first outlined by Wallas in 1926 (*The Art of Thought*, New York: Harcourt, Brace).

2. Charles Nicolle, "The Mechanism and Transmission of Typhus," in Rene Taton, *Reason and Chance in Scientific Discovery* (New York: Philosophical Library, 1957), pp. 76–78. Reprinted by permission of Philosophical Library, Inc.

3. Taton, p. 85.

4. Taton, p. 88.

2

♦ ♦ ♦

Library Resources

Students' experience researching in libraries varies from expert to novice. Your own experience will fall somewhere along that continuum. But even if you have used libraries before, there is always more to learn. College libraries are typically large, complex entities that often seem to have a life of their own. They are constantly changing as new information and methods to access information are incorporated by librarians.

Your library contains many important general research tools with which you need to become familiar to conduct any research project. Also in your library are specific resources that are important for research in particular disciplines. This chapter covers information on general library resources in the sciences and technology. You may find that some of this chapter is review, but you will certainly also encounter sources of which you were not previously aware.

THE LIBRARY REFERENCE AREA

Most library research begins in the library reference area. Reference librarians are excellent resources when you need help finding information in the library. However, you need to know enough about libraries to ask the librarian to help you, just as you need to know enough about your car to suggest to a mechanic where to begin when your car needs repair. It is not productive to walk up to a mechanic and simply ask for help. You need to explain the specifics of your particular problem and describe the make, model, age, and condition of your car. Similarly, you need to tell a librarian what kind of project you are working on, what information you need, and in what form that information is likely to be stored. So that you can ask the right questions, you must first take the time to learn your way around the library. Working your way through this chapter is a good place to begin.

Librarians use terminology that you should become familiar with. Particular terms are defined throughout this chapter. We will start by looking at some major library sources and what distinguishes them.

In the heading of this section of the chapter you saw the word *reference*. As you might suspect, a reference is something that *refers* to something else. In your library there is an area designated as the reference area and a person called the reference librarian. The reference area may be a separate room or simply a section of the library. In the reference area you will find books containing brief factual answers to such questions as, What is the meaning of a particular word? What is the population density of a particular state? What is the birthplace of a certain famous person? Also in the reference area are books that refer you to other sources. These library tools—bibliographies and indexes—will help you find particular articles written about particular subjects. The reference area of the library is usually the place to begin any research project. Reference sources include the following general types of works:

Abstracts: Short summaries of larger works; may be included in an index

Almanacs: Compendia of useful and interesting facts on specific subjects

Atlases: Bound volumes of maps, charts, or tables illustrating a specific subject

Bibliographies: Lists of books or articles about particular related subjects

Biographies: Works that provide information on the life and writings of famous people, living and dead

Dictionaries: Works that provide information about words, such as meaning, spelling, usage, pronunciation

Encyclopedias: Works that provide concise overviews of topics, including people, places, ideas, subject areas

Handbooks: Books of instruction, guidance, or information of a general nature

Indexes: Books or parts of books that point to where information can be found, such as in journals, magazines, newspapers, or books

Reviews: Works that analyze and comment on other works, such as films, novels, plays, or even research

Serials: Periodicals (magazines) that are published at specific intervals and professional (scholarly) journals that contain articles and research reports in a specific field

ONLINE COMPUTER CATALOGS

One of the most visible computer tools in libraries today is the online catalog. Such a system, which supplements or replaces the traditional card catalog, is designed to provide library materials service, such as circulation, cataloging, and location of materials within the library collection. Most online catalogs are searchable by author, title, and subject (as are card catalogs), but many are also searchable by keyword or by combinations of subjects or keywords.

You should become familiar with your library's online catalog as soon as possible. (Note: Some online catalogs cover only the most recent additions to the library's collection, in which case older books still may be found by looking in the card catalog.) Check with your librarian to find out exactly what materials (e.g., books, magazines, journals, government documents) are cataloged in your library's online catalog.

How to Search the Online Catalog

When you log-on to your library's online catalog, you will typically see a menu of choices listing the various databases available for searching. For example, the library at my university includes the following databases in its online catalog, all searchable from the same computer. To enter any one of these databases, you make the appropriate selection from the main menu.

General Book Collection (OPAC, Online Public Access Catalog) lists the books, government documents, and audiovisual materials in the library.

General Periodicals Index (WRGA, Wilson Readers' Guide to Periodical Literature) lists articles from popular magazines and journals and corresponds to the *Readers' Guide to Periodical Literature* from 1985 to the present.

To search for articles written in particular disciplines, our library also provides several specialized databases in the online catalog:

Wilson Guide to Business Periodicals, Education Index, and the Social Sciences Index (WSOC) indexes journal articles in business, education, and social sciences.
Wilson Guide to Art Index and Humanities Index (WHUM) indexes journal articles in humanities and arts.
Wilson Guide to Applied Science and Technology Index, Biological and Agricultural Sciences Index, and General Sciences Index (WSCI) indexes journal articles in sciences, agriculture, and technology.
Current Contents Articles and Journals (CART and CJOU) indexes the table of contents for recent journal issues in several fields of the sciences and social sciences (Figure 2-1).

You will need to know enough about how information in your library is organized to be able to select the appropriate databases from the computer menu. Many libraries offer instruction in the use of their online catalog. If yours does not, plan to spend the time you need to become a confident user of your library's computer system. Take advantage of online help menus provided by the computer, and be sure to ask a librarian for help when stumped. Librarians are trained to be "user friendly."

Internet

Many libraries provide access to information in other libraries through Internet services via computer. The Internet is a computer "network of networks" from campuses as well as regional, national, and even international sources that are joined into a single network. It is like a worldwide network of information highways with all the freeways and byways connected. Internet in our library offers an opportunity to explore other Utah library catalogs, for example, or to find information through many different lists or "gophers" that store information. Students can use Internet to find information on sample job interview questions offered in a gopher by USU's career services. Or, you can locate the most recent White House press release on the budget.

The main challenge right now is learning how to "mine" the Internet, to tap into its vast array of information. New systems to help consumers find their way through the maze on the information superhighway are constantly being developed. There are, for example, online services such as America Online that provide easy-to-use interfaces, allowing access to hundreds of resources. There are also navigational helpers such as the World Wide Web, which have established standards and protocols to make searching among a variety of networks virtually seamless.

```
              WELCOME
        UTAH STATE UNIVERSITY
           MERRILL LIBRARY

To select a database, type appropriate four letter
code, i.e. OPAC and press ENTER. To return to this
menu screen, type START.

                MERLIN
OPAC            USU LIBRARY COLLECTION
                JOURNAL/PERIODICAL INDEXES
*WRGA           POPULAR MAGAZINE INDEX
*WHUM           HUMANITIES & ART INDEXES
*WSCI           AGRICULTURE/BIOLOGY & SCIENCE
                TECHNOLOGY
*WSOC           BUSINESS/EDUCATION & SOCIAL SCIENCES
                CURRENT CONTENTS
*CART           CURRENT CONTENTS ARTICLES
*CJOU           CURRENT CONTENTS JOURNALS
            * Databases that require Sign-On.
---------------------Page 1 of 1------------------
HELP  Select a database label from above
NEWs (Library System News)


Database Selection:
Alt-Z FOR HELP₃ VT100   ₃ FDX ₃ 19200 N81 ₃ LOG
CLOSED ₃ PRINT OFF ₃ OFF-LINE
```

FIGURE 2-1 Main Menu

The most popular interfaces to information on the Internet are Mosaic and Netscape. Mosaic was developed by the National Center for Supercomputing Applications (NCSA) in Champaign, Illinois. The Mosaic and Netscape client/server applications are designed so that information can be distributed and retrieved over the Internet using a graphical user interface, thus making it possible for clients to locate other servers of different types along the Internet while maintaining the same consistent graphics on the client's screen. Through hypertext links, users can move easily about the Internet by simply pointing and clicking on Mosaic or Netscape graphics. The Internet is in constant flux and growing at a phenomenal rate, so you would be wise to explore the Internet resources available on your particular campus. The best way to learn about Internet is to practice using it yourself.

Subject Headings

When using online catalogs, it is important to become familiar with the *Library of Congress Subject Headings* (*LCSH*), a listing of subject areas using specific terminology. To search the database by subject, you must provide the computer with exact subject headings, those officially used by the Library of Congress as listed in the *LCSH*. However, sometimes it is difficult to predict which terms will be used. The *LCSH* is your key or guide to the subjects recognized by the online catalog: it lists the subject headings that are assigned to books and materials in the library's collection. As there is often more than one way to describe a topic, the *LCSH* gives the exact format (wording and punctuation) for subject headings as they will appear in the database.

Before beginning to search the online catalog, first check the *LCSH* to become familiar with all the possible subject terms related to your topic. Keep a comprehensive list of all the subject headings that you are using in your search of the library's holdings. These subject headings will be useful to you not only for the library's book collection, but also when you begin searching for articles in magazines and journals.

Title and Author Searches The computer can search quickly through its database when you provide it with a correct title or an author's name (T = Title; A = Author). If there is more than one work by an author, the computer lists them all, and then allows you to select the one you are looking for. You may need to request an expanded screen (or "long view") with more details about the specific work you have selected. The computer should also let you know whether or not the book has been checked out of the library and provide you with a call number to help you locate the book. In some libraries, it is possible to place a "hold" on a work through the computer, with a provision that you will be notified as soon as the book is back in circulation (Figure 2-2).

Subject Searches To search by subject, once again be certain to use the exact *LCSH* subject heading as found in the *LCSH* volumes. Tell the computer that you intend to search by subject by entering the appropriate command along with the exact subject heading (in our library, for example, you type "S = Subject"). The computer then tells you how many items in its database are cataloged under that subject head-

```
Search Request: T=CURRENT ORNITHOLOGY                        USU
SERIAL - Record 1 of 1 Entry Found
-----------------------------------------------------------------
Title:              Current ornithology.

Published:          New York : Plenum Press, c1983-
                    Vol. 1-

Frequency:          Annual

Subjects:           Ornithology.
                    Ornithological research.

Other authors:      Johnston, Richard F.
-----------------------------------------------------------------
LOCATION:           CALL NUMBER:     STATUS:
SciTech LIBRARY     QL671 .C87x      Enter HOL 1 for holdings
STACKS-4th FLOOR
--------------------------------------  Page 1 of 1  ---------
STArt over          HOLdings
HELp                BRIef view
OTHer options

NEXT COMMAND:
~library
```

FIGURE 2-2   Online Title Search

ing (called the number of "hits") and provides you with a list of titles (Figure 2-3 on page 18).

If the computer tells you that there are several hundred books with your subject heading, you may need to narrow your subject search. The *LCSH* will provide you with narrower terms (labeled "NT"). Or you can narrow your search by combining key words (see Keyword Searching section in this chapter). On the other hand, if the computer tells you there are only one or two titles with your subject heading, you may need to broaden your subject search. Again, the *LCSH* will suggest related terms (labeled "RT") or broader terms ("BT") that you can try as well.

Call Number Searching.   This is an excellent way to locate materials similar to works for which you currently have a call number. By typing in the exact call number of a known work, you can ask the computer to list all of the call numbers that come before and after that work in the computer's memory. This means that it is possible to read titles on the computer screen by call number, just as you would browse a shelf in the library and retrieve other books found on surrounding shelves.

Library Resources

```
Search Request:   S=ORNITHOLOGY                         USU CATALOG
Search Results:   59 Entries Found
-----------------------------------------------------------------
    ORNITHOLOGY
 1  ABSTRACTS COOPER ORNITHOLOGICAL SOCIETY FIFT <1982>
 2  AVIAN BIOLOGY <1971>
 3  BIOLOGY AND COMPARATIVE PHYSIOLOGY OF BIRDS <1960>
 4  BIRD ITS LIFE AND STRUCTURE <1951>
 5  BOOK OF BIRD LIFE A STUDY OF BIRDS IN THEIR <1961>
 6  CURRENT ORNITHOLOGY <NEW YORK> serial
 7  FUNDAMENTALS OF ORNITHOLOGY <1959>
 8  FUNDAMENTALS OF ORNITHOLOGY <1976>
 9  INSTRUCTIONS TO YOUNG ORNITHOLOGISTS <1959>
10  INSTRUCTIONS TO YOUNG ORNITHOLOGISTS BIRD BI <1959>
11  INTRODUCTION TO ORNITHOLOGY <1955>
12  INTRODUCTION TO ORNITHOLOGY <1963>
13  INTRODUCTION TO ORNITHOLOGY <1975>
14  LABORATORY AND FIELD MANUAL OF ORNITHOLOGY <1956>
------------------------------------- CONTINUED on next page
STArt over        Type number to display record      FORward page
HELp              GUIde
OTHer options

NEXT COMMAND:
~library
```

FIGURE 2-3 Online Subject Search

Keyword Searching. Keyword searching allows for searching under the most important term or terms for your research project that you have identified. The computer locates items in its database that use a particular keyword anywhere in a work's record. However, if the record doesn't happen to include that particular keyword or term, the computer typically will not be able to supply the synonym. Therefore, you still need to try as many keywords and subject headings as you can think of in a database search. For example, if you are searching the subject UFOs, you might use "UFO" as a keyword. But you might also want to try other keywords, such as "flying saucers" or "paranormal events."

Advanced Keyword Searching. Using keywords, it is possible to perform combined searches of two or more terms. Combining keywords helps to limit an otherwise broad topic. For example, if you were interested in teenage pregnancy but only for the state of California, you could combine the terms "teenage pregnancy" and "California" to narrow your search, thus searching the collection only for those sources that include both keywords. This kind of focused searching offers distinct advantages over using a card catalog. If your library offers in-

```
Search Request: K=FALCONS                                  USU CATALOG
Search Results: 41 Entries Found                          Keyword Index
------------------------------------------------------------------------
    DATE TITLE:                                      AUTHOR:
 1  1994 City peregrine : a ten year saga of New Yo Frank, Saul
 2  1993 Fat content of American kestrels (Falco sp Harden, Shari M
 3  1993 Merlin : Falco columbarius
 4  1993 The Vertical environment <visual>
 5  1992 Falcons of Arabia <visual>
 6  1992 Flight of the falcon                       Tennesen, Michael
 7  1992 The hunting falcon                         Haak, Bruce A
 8  1992 Peregrine falcons                          Savage, Candace She
 9  1992 Wings for my flight : the peregrine falcon Houle, Marcy Cottre
10  1991 The compleat falconer                      Beebe, Frank L
11  1991 Romancing the falcon                       Weiss, Steve
12  1990 Hawking the American West : romancing the  Weiss, Steve
13  1990 Hawks, eagles & falcons of North America : Johnsgard, Paul A
14  1990 On the far side of the mountain            George, Jean Craigh
------------------------------------------------------------------------
STArt over       Type number to display record           FORward page
HELp
OTHer options

NEXT COMMAND:
~library
```

FIGURE 2-4 Online Keyword Search

struction in advanced database searching, you would be wise to take the opportunity to learn the "tricks of the trade" (Figure 2-4).

Recording Bibliographic Information

Once you have a listing of books from the online catalog, you need to note each book's specific call number so that you can locate it in your library's collection. It may be possible for you to send a print command from your online catalog to request a printout of your search results. If not, you will need to be certain you write down the titles, along with their complete bibliographic information, on your working bibliography in your research notebook. Complete bibliographic information includes:

  Author(s) full name, including initials
  Title of the book, including subtitles and editions
  Place (city and state) where the book was published
  Name of the book's publisher
  Date of publication

MICROFORMS

A few libraries organize their holdings on microforms rather than computer databases or catalog cards. Microforms are either microfilm or microfiche; both make use of methods to condense information in a very compact form. Machines called microform readers are necessary to read either form. Information stored on microform is similar to that found on traditional cards. If your library uses this system, check with the reference librarian for instructions about using the microform readers.

LOCATING ARTICLES IN SERIALS: POPULAR PERIODICALS

Periodicals are popular magazines and newspapers printed at regular intervals, such as daily, weekly, monthly, or quarterly. Typically, when researching for upper-division science courses, you would not use popular magazines at all, but rather would use the scholarly journals discussed below. However, when you are engaged in a more general search you may wish to use one of these indexes of general-interest periodicals:

Magazine Index. Los Altos, CA.: Information Access, 1978 to present.
Readers' Guide to Periodical Literature. New York: Wilson, 1901 to present.

Newspapers

Your library probably stores back issues of newspapers on microform. To gain access to articles in the *New York Times,* use the *New York Times Index,* which lists all major articles from the *Times* from 1913 to the present. The *Newspaper Index* lists articles from the *Chicago Tribune, Los Angeles Times, New Orleans Times-Picayune,* and *Washington Post.* Both indexes are arranged by subject. For business news, use the *Wall Street Journal Index.* Newspaper indexes may be available in your library in both print and database forms.

Your library has a particular system for listing magazine and newspaper holdings. This serials listing may appear in the online catalog, in a separate catalog, in the main card catalog, on computer printouts, or on microform. Consult your librarian to determine which system your library uses. The serials listing tells you to which periodicals your library subscribes, where they are located in the library, the inclusive dates of the issues your library has, and whether the issues are in bound or unbound volumes or on microform.

Once you have obtained the call number of the magazine containing your article from the serials listing, you will be able to find the magazine itself, whether it is in a bound volume, in the current-periodicals section of the library, or on microform. Once you have the article in hand, be sure to copy down the complete publication information, which is often only abbreviated in the index. Writing everything down at this point will prevent your having to return to the library later for information you neglected to note originally.

When researching your topic, write down or print out the *complete* citation of any relevant article. The citation includes:

>   the author (if given)
>   the title of the article
>   the name of the magazine
>   the volume number
>   the issue number
>   the date published
>   the inclusive pages of the article

LOCATING ARTICLES IN SERIALS: PROFESSIONAL JOURNALS

When you are researching a technical subject, you want to refer to articles written on the subject by professionals in the field. Professional journal articles, sometimes called serials because they are printed in series, are indexed in much the same way as the general periodicals previously discussed. However, numerous specialized indexes and databases exist for professional articles, and each index or database covers a particular discipline or subject area. (See "Discipline-Specific Resources" on page 26.)

Once you have located an appropriate index, begin looking up your topic in the most recent volume first. If you were researching the topic "drug abuse," Figure 2-5 on page 22 provides an example of what you might find in the *Social Sciences Index*.

For articles that sound promising for your research, copy down the complete citation. To find the full version of abbreviated journal names, turn to "Abbreviations of Periodicals Indexed," usually at the front of the index volume. For example, you would find that *Congr Q Wkly Rep* is the abbreviation for *Congressional Quarterly Weekly Reports*. Then locate the particular journal by using your library's listing of ser-

Hawks
>*See also*
>Falcons
>Kites (Birds)

Hawk watching in New York. L. Chamberlaine. il *Conservationist* 47:42-7 N/D '92

The influence of weather on hawk movements in coastal northern California. L. S. Hall and others. bibl il *Wilson Bull* 104:447-61 S '92

Pitching equilibrium, wing span and tail span in a gliding Harris' hawk, Parabuteo unicinctus. V. A. Tucker. bibl il *J Exp Biol* 165:21-41 Ap '92

Status, nesting density, and macrohabitat selection of red-shouldered hawks in northern New Jersey. T. Bosakowski and others. bibl il *Wilson Bull* 104:434-46 S '92

Hay fever

John Bostock, hay fever, and the mechanism of allergy. R. Finn. bibl *Lancet* 340:1453-5 D 12 '92

>**Therapy**

Hay fever: antihistamine premedication may reduce symptoms. *Geriatrics* 47:25 S '92

Rhinitis and asthma. M. Kaliner and R. Lemanske. bibl il *J Am Med Assoc* 268:2807-29 N 25 '92

Seasonal allergic rhinitis responds to once-daily nasal steroid. S. Stroud and J. Dyer. *Am J Nurs* 92:47 Jl '92

Hazardous substances
>*See also*
>Chemical spills
>Poisonous gases

The asbestos mess [noting the Amphibole Hypothesis] T. Harris. il *Garbage* 4:44-9 D '92/Ja '93

Dangerous technology dumped on Third World. I. Anderson. *New Sci* 133:9 Mr 7 '92

Emerging issues in health care: the role of the environmental health residency. G. Byrns and others. bibl il *J Environ Health* 55:31-5 Jl/Ag '92

FIGURE 2-5 General Science Index

---

ial holdings, just as you did for magazines and newspapers. Professional journals are stored in bound or unbound volumes (the latter for very recent editions) or on microform. The citation you have obtained from the index is your key to finding a particular article. Thus, it is crucial that you copy down the citation information accurately and completely.

Many of the indexes to scholarly journals follow the format of the *General Science Index* (see Figure 2-5). However, there are sometimes variations in the index formats. Take the time to get acquainted with the arrangement of each index by reading the explanation in the index's preface. If you are still confused about how an index is organized, ask your reference librarian for help. As mentioned before, many indexes are now contained on databases so that they may be searched with computers. Check with your librarian to discover which indexes in your library may be searched on computer.

LOCATING GOVERNMENT DOCUMENTS

The U.S. government is one of the largest publishers of information and is a rich source of materials in almost every field, from aeronautics to zoology. Government documents are sometimes listed in a separate database or catalog from the main online or card catalog. Check with your reference librarian to discern how your library catalogs its documents.

The *Monthly Catalog of United States Government Publications* is the comprehensive bibliography that lists all publications received by the Government Printing Office for printing and distribution. The *Monthly Catalog,* established in 1895, is the best overall guide to finding government sources. At the back of each monthly register are indexes that provide access to the documents by authors, subjects, and series/report numbers. The monthly indexes are cumulated (that is, brought together into one volume) both semiannually and annually for ease of access.

Other ways to find government documents include the following indexes:

Index to U.S. Government Periodicals
Public Affairs Information Services (PAIS)
Resources in Education (RIE)
U.S. Government Reports, Announcements and Index
(NTIS—National Technical Information Service)

The *Index to U.S. Government Periodicals* provides access by author and subject to 180 government periodicals. The *PAIS* lists by subject current books, pamphlets, periodical articles, and government publications in the field of economics and public affairs. *RIE* lists government-sponsored reports related to the field of education. The *U.S. Government Reports, Announcements and Index (NTIS)* lists government-sponsored research in the technical sciences by subject fields: aeronautics; agriculture; astronomy and astrophysics; atmospheric sciences; behavioral and social sciences; biological and medical sciences; chemistry; earth science and oceanography; electronic and electrical engineering; energy conversion; materials; mathematical sciences; mechanical, industrial, civil, and marine engineering; methods and equipment; military science; navigation, communications, detection; nuclear science and technology; physics; and propulsion and fuels.

OTHER TECHNOLOGIES

Online Database Services

To supplement the searches you have been able to conduct in your own library, online database searches may be conducted for you by trained search librarians who have access through a telecommunication link to a central databasing service such as DIALOG. When you want to conduct such a search, make an appointment with a librarian to discuss your search needs, possible databases, and likely subject headings. The librarian then conducts the search and provides you with a printout of sources or abstracts. The cost of the search is typically passed on to you, so you should discuss any search with a trained librarian first to decide whether it is feasible and economical. (See Figure 2-6.)

CD-ROM Searching

Many libraries are now providing patrons with the opportunity to search databases themselves, using microcomputers connected to compact disk units (CD-ROMs). With compact disk technology, large databases can be made accessible and easy to use. For example, the education index ERIC (Education Research Information Clearinghouse) is now available in this format, as is the business index ABI-INFORM. Check your library reference area to see whether such tools are available to you. These databases are subject specific, so you need to find out just which journals or subjects they index. But they often can provide a quick alternative to searching the print indexes on your subject. Another bonus of using CD-ROMs is that the information found in your search can usually be sent to a printer or downloaded onto your own computer disk.

As with any computerized search, it is important to know your subject headings and terminology. Many of the computer databases use their own "controlled vocabulary," which may vary slightly from the subject headings listed in the *LCSH*. Check with your librarian to discern whether there is a "thesaurus" or listing of subject headings for the particular database you are using. ERIC on-disk, for example, uses the "ERIC Descriptors" as its method of cataloging by subject.

```
PRINTS User:U0020650 28Sep84 PRINT 13/5/1-117
```

Database service date — DIALOG File 1: ERIC - 66-84/Sep

```
EJ299865    HE518071
   Many Colleges Limit Students' Use of Central Computers for
Writing.
   Turner, Judith Axler
   CAUSE/EFFECT, v7 n3 p6-7 May 1984
   Available from: UMI
   Language: English
   Document Type: POSITION PAPER (120); PROJECT DESCRIPTION (141)
   Journal Announcement: CIJSEP84
   Allocating limited resources to an unlimited demand is an
issue faced by data processing management in higher education.
Use of the central computer for word processing is creating a
demand at many institutions that is stretching and exceeding the
available computing resources. (Author/MLW)
   Descriptors: *College Students; *Computers; Data Processing;
Higher Education; *Time Management; Use Studies; *Word Process-
ing; *Writing (Composition)
   Identifiers: *Computer Centers; Yale University CT
```

Education Journal (EJ) identification number — EJ298427   IR512505 ←*Title*
```
   A Dyslexic Can Compose on a Computer.
   Arms, Valarie M.                                             ←Author
   Educational Technology, v24 n1 p39-41 Jan 1984   ←Journal
   Available from: UMI
   Language: English                                            publication
   Document Type: PROJECT DESCRIPTION (141)         data
   Journal Announcement: CIJAUG84
```
Abstract —
```
   Describes the strategies used by a technical writing teacher
who encouraged a dyslexic university engineering student to use
a microcomputer as an aid in composition writing, and discusses
how a word processing program was used to make the writing
process easier and increase the student's self-confidence. (MBR)
```
Subject headings — Descriptors:
```
*Dyslexia; Higher Education; Learning Disabilities; Learning Mo-
tivation; *Microcomputers; *Teaching Methods; *Word Processing;
*Writing (Composition)

EJ298270    FL515801
   Computer-Assisted Text-Analysis for ESL Students.
   Reid, Joy; And Others
   CALICO Journal, v1 n3 p40-42 Dec 1983
   Language: English
   Document Type: PROJECT DESCRIPTION (141); POSITION PAPER (120)
   Journal Announcement: CIJAUG84
   Reports an investigation into possibilities of using word
processors and text analysis software with English as second
language (ESL) students to determine (1) if foreign students can
learn to use computer equipment, (2) if students feel time in-
vested is worthwhile, and (3) if ESL students' problems with
PAGE:2
                                              Item 1 of 117
writing American academic prose can be remedied by this type of
assistance. (SL)
   Descriptors: Comparative Analysis; *Computer Assisted In-
struction; *English (Second Language); Grammar; Modern Language
Curriculum; Second Language Learning; *Student Participation;
Vocabulary Development; *Word Processing; *Writing Skills

EJ298267    FL515798
   Computer-Assisted Language Learning at the University of
Dundee.
   Lewis, Derek R.
   CALICO Journal, v1 n3 p10-12 Dec 1983
   Language: English
   Document Type: NON-CLASSROOM MATERIAL (055); POSITION PAPER
(120)
   Journal Announcement: CIJAUG84
   Presents an overview of activities in field of computer-as-
sisted language learning at the University of Dundee (Scotland).
These include: (1) use and testing of a self-instructional
teacher's package, (2) development of the computer-controlled
```

FIGURE 2-6 Online Database

ACCESSING INFORMATION FROM OTHER LIBRARIES

If, in your search, you discover that some item you need is not located in your library, it is still possible to find the item in another library. An online computer database called the OCLC (Online Computer Library Center) provides thousands of libraries with connections to each other's catalogs. By searching the OCLC database (by author or title), you can quickly ascertain if a nearby library contains the needed item; then you may request it through interlibrary loan or secure it yourself.

DISCIPLINE-SPECIFIC RESOURCES FOR SCIENCES AND TECHNOLOGY

General Sources and Guides to Literature

Guide to Sources for Agricultural and Biological Research. J. Blanchard and L. Farrell. Los Angeles: University of California Press, 1981.

Information Sources in Agriculture and Horticulture, 2nd ed. G. P. Lilley. UK: H. Zell, 1992.

Information Sources in Engineering. 2nd ed. L. Anthony, ed. UK: H. Zell, 1985.

Information Sources in Physics. 2nd ed. D. Shaw. UK: H. Zell, 1985.

Information Sources in the Life Sciences. 4th ed. H. V. Wyatt, ed. UK: H. Zell, 1992.

Information Sources in the Medical Sciences. 4th ed. L. T. Morton and S. Godbolt, eds. UK: Bowker-Saur, 1992. A useful reference guide for all medical fields.

Information Sources in Science and Technology. C. D. Hurt. Littleton, CO: Libraries Unlimited, 1988. A general reference guide for the sciences.

Reference Sources in Science and Technology. E. J. Lamsworth. Metuchen, NJ: Scarecrow, 1972.

Science and Engineering Sourcebook. Cass R. Lewart. Littleton, CO: Libraries Unlimited, 1982. Lists and annotates reference works in various scientific fields.

Chambers Dictionary of Earth Science. Peter Walker, ed. Edinburgh: W. & R. Chambers, 1992. Concise source for definitions of terms.

Dictionary of Chemistry. Sybil P. Parker, ed. New York: McGraw-Hill, 1985.

Dictionary of Artificial Intelligence and Robotics. Jerry Rosenberg. New York: Wiley, 1986.

Dictionary of the Biological Sciences. P. Gray. Melbourne, FL: Krieger, 1982.

Dictionary of Computing. 3rd ed. New York: Oxford University Press, 1991. Lists over 4,000 terms used in computing and associated fields of electronics, mathematics, and logic.

Dictionary of Geology and Geophysics. D. F. Lapidus. New York: Facts on File Publications, 1987. Defines many terms in the context of modern geological theories.

Dictionary of Electrical and Electronic Engineering. New York: McGraw-Hill, 1985. Very comprehensive guide to terms.

Dictionary of Inventions and Discoveries. 2nd ed. E. F. Carter. New York: Crane Russak, 1976. Catalogs and describes major scientific inventions and discoveries.

Dictionary of Physics. New York: McGraw-Hill, 1986. Comprehensive dictionary of terms.

Glossary of Chemical Terms. 2nd ed. C. A. Hampel and G. G. Hawley. New York: Van Nostrand Reinhold, 1982. Helpful definitions.

McGraw-Hill Dictionary of Earth Sciences. 3rd ed. New York: McGraw-Hill, 1984. Helpful definitions.

McGraw-Hill Dictionary of Scientific and Technical Terms. 4th ed. New York: McGraw-Hill, 1989. Provides clear definitions of terminology.

A Modern Dictionary of Geography. 2nd ed. J. Small and M. Witherick. Baltimore: Edward Arnold, 1989. Provides definitions of terms that are accessible to college students.

Handbooks, Atlases, and Almanacs

Handbook of Chemistry and Physics. 58th ed. Cleveland: Chemical Rubber, 1913–present. Provides facts and data on chemistry and physics.

Materials Handbook. 13th ed. G. S. Brady. New York: McGraw-Hill, 1991. Describes nature and properties of commercially available materials.

Medical and Health Information Directory: Organizations, Agencies, and Institutions. 7th ed. Detroit: Gale, 1994. Comprehensive guidebook.

Physician's Handbook. 21st ed. M. A. Krupp et al. E. Norwalk, CT: Appleton and Lange, 1986. Useful, quick reference book for all medical questions.

Standard Handbook for Civil Engineers. 3rd ed. F. S. Merritt, ed. New York: McGraw-Hill, 1983. Provides basic information in an easy reference format.

Standard Handbook for Electrical Engineers. 12th ed. D. G. Fink and H. W. Beaty, eds. New York: McGraw-Hill, 1987. Provides basic information in an easy reference format.

Encyclopedias

The Astronomy Encyclopedia. P. Moore, ed. Stafford, England: Mitchell Beagley, 1989. Illustrated.

Cambridge Encyclopedia of Earth Sciences. D. G. Smith, ed. Cambridge, England: Cambridge University Press, 1982. Short articles of interest to geologists, geographers, and so on.

Cambridge Encyclopedia of Life Sciences. A. Friday and D. S. Ingram, eds. Cambridge, England: Cambridge University Press, 1985. Short articles of interest to biologists, zoologists, and others.

Classification and Synopsis of Living Organisms. S. P. Parker, ed. New York: McGraw-Hill, 1982. Invaluable reference tool for life sciences.

The Encyclopedia of Astronomy and Astrophysics. Maran, ed. New York: Van Nostrand Reinhold, 1991. Concise summaries of information geared for a nontechnical audience.

The Encyclopedia of Bioethics. W. Reich, ed. New York: Macmillan, 1982.

Encyclopedia of Computer Science and Engineering, revised ed. A. Ralston, ed. New York: Van Nostrand Reinhold, 1992. Provides concise information in the fields of computer science and engineering.

Encyclopedia of Computer Science and Technology. J. Belzer, ed. New York: Dekker, 1990. Short articles on subjects in computer science.

Encyclopedia of Physical Science and Technology. 15 vols. R. A. Meyers, ed. Orlando: Academic Press, 1987. A comprehensive encyclopedia on the status of knowledge across the entire field of physical science and related technologies.

The Encyclopedia of Physics. R. G. Lerner and G. L. Trigg, eds. Reading, MA: Addison-Wesley, 1980. Provides background information on major principles and problems in physics.

Grzimek's Animal Life Encyclopedia. B. Grzimek, ed. New York: Van Nostrand Reinhold, 1972-present. Provides an overview of the animal kingdom, with illustrations.

Grzimek's Encyclopedia of Mammals. B. Grzimek, ed. New York: MacGraw-Hill, 1990. Provides general information on the study of mammals.

McGraw-Hill Encyclopedia of Energy. 2nd ed. New York: McGraw-Hill, 1980. Short articles about all fields of energy.

McGraw-Hill Encyclopedia of Environmental Science. 2nd ed. S. P. Parker, ed. New York: McGraw-Hill, 1980. Provides information on the earth's resources and how they have been used.

McGraw-Hill Encyclopedia of Science and Technology. 15 vols. 7th ed. New York: McGraw-Hill, 1992. Provides concise, current background information on scientific and technical topics; an excellent place to begin a science research project, since the articles are not written for specialists.

McGraw-Hill Yearbook of Science and Technology. New York: McGraw-Hill, annual. Updates the encyclopedia (listed above) every year. Consult the yearbook for the most recent developments in a particular field.

Van Nostrand's Scientific Encyclopedia. 7th ed. New York: Van Nostrand Reinhold, 1988. Provides concise background information on a variety of scientific disciplines.

VNR Concise Encyclopedia of Mathematics. W. Gellert et al., eds. New York: Van Nostrand Reinhold, 1989. Short articles on all areas of mathematics.

VNR Encyclopedia of Chemistry. 4th ed. D. M. Considine, ed. New York: Van Nostrand Reinhold, 1984. Short articles on all areas of chemistry.

Biographies

American Men and Women of Science. 16th ed. J. Cattell, ed. New York: Bowker, 1986. Provides information on living, active scientists in the fields of economics, sociology, political science, statistics, psychology, geography, and anthropology.

Dictionary of Scientific Biography. New York: Scribner's, 1970–1981. Provides information on scientists from classical to modern times. Covers only scientists who are no longer living.

National Academy of Sciences, Biographical Memoirs. Washington, D.C.: National Academy of Sciences, 1877–present. Provides information on American scientists.

Who's Who in Science in Europe. 5th ed. Detroit: Gale, 1984.

Indexes and Abstracts

AEROSPACE

Aerospace Medicine and Biology
*International Aerospace Abstracts
*Scientific and Technical Aerospace Reports
*U.S. Government Reports, Announcements and Index (NTIS)

AGRICULTURE

*Agricola
*Agricultural Engineering Abstracts
Agritrop
*Agronomy Abstracts
*Bibliography of Agriculture
*Biological and Agricultural Index
FAO Documentation, Government Documents Index
Farm and Garden Index
*Fertilizer Abstracts
*Field Crop Abstracts
*Herbage Abstracts
*Seed Abstracts
*Soils and Fertilizers
*World Agricultural Economics and Rural Sociology Abstracts

ANIMAL SCIENCE

*Animal Behavior Abstracts
*Animal Breeding Abstracts
Bibliography of Reproduction
*Dairy Science Abstracts
*Index Veterinarius
*Veterinary Bulletin

ASTRONOMY

*Astronomy and Astrophysics Abstracts
*Meteorological and Geoastrophysical Abstracts

BIOLOGY, BOTANY, ENTOMOLOGY, AND ZOOLOGY

Asher's Guide to Botanical Periodicals

*Biological Abstracts
*Biological and Agricultural Index
Botanical Abstracts
Biology Digest
Current Advances in Plant Science
*Entomology Abstracts
**Genetics Abstracts
*Horticultural Abstracts
International Abstracts of Biological Science
*Plant Breeding Abstracts
*Review of Applied Entomology
*Review of Medical and Veterinary Entomology
Review of Plant Pathology
*Soils and Fertilizers
Torrey Botanical Club Bulletin
*Virology Abstracts
*Weed Abstracts
*Zoological Record

CHEMISTRY

Analytical Abstracts
*Chemical Abstracts

COMPUTERS AND ROBOTICS

*Artificial Intelligence Abstracts
*CAD/CAM Abstracts Index
*Computer Abstracts
*Computer and Control Abstracts
Computing Reviews
Data Processing Digest
*Electrical and Electronic Abstracts
*Microcomputer Index
Robomatix Reporter

ENERGY AND PHYSICS

Energy Abstracts for Policy Analysis
*Energy Information Abstracts
*Energy Research Abstracts
INS Atomindex
Nuclear Science Abstracts
*Physics Abstracts

* = computer searching available

ENGINEERING

Agricultural Engineering Abstracts
Applied Mechanical Reviews
Civil Engineering Hydraulics Abstracts
*Electrical and Electronics Abstracts
*Engineering Index
International Aerospace Abstracts
*ISMEC Bulletin (mechanical engineering)

ENVIRONMENT AND ECOLOGY

Abstracts on Health Effects of Environmental Pollutants
Air Pollution Abstracts
Current Advances in Ecological Sciences
Ecological Abstracts
Ecology Abstracts
*Environment Abstracts
*Environment Index
Environment Information Access
*Environmental Periodicals Bibliography
*Pollution Abstracts
*Selected Water Resource Abstracts
*Water Resources Abstracts

FOOD SCIENCE AND NUTRITION

*Food Science and Technology Abstracts
*Foods Adlibra
*Nutrition Abstracts and Reviews
Nutrition Planning

FORESTRY

Fire Technology Abstracts
*Forestry Abstracts

GEOGRAPHY, GEOLOGY, AND MINING

*Bibliography and Index of Geology
*Bibliography of North American Geology
Current Geographical Publications
Deep Sea Research Part B
*GEO Abstracts

*Geographical Abstracts
Geophysical Abstracts
*Meteorological and Geoastrophysical Abstracts
*Oceanic Abstracts
Population Index

MATHEMATICS AND STATISTICS

*American Statistics Index
Current Mathematical Publications
Demographic Yearbook
*Mathematical Reviews
Statistical Abstract of the United States
Statistical Reference Index
Statistical Theory and Method Abstracts
Statistical Yearbook

MEDICINE, NURSING, AND ALLIED HEALTH FIELDS

AIDS Bibliography
*Ageline
Bibliography of Reproduction
*Cumulated Index Medicus
*Cumulative Index to Nursing and Allied Health Literature
Endocrinology Index
*International Nursing Index
*Medline Clinical Collection

MEDOC

*Physical Fitness/Sports Medicine
Virology Abstracts

SCIENCE—GENERAL

*Applied Science and Technology Index
Current Bibliographic Directory of the Arts and Sciences
*General Science Index
Index to Scientific and Technical Proceedings
*Science Citation Index

TEXTILES

Clothing and Textile Arts Index

Clothing Index
Textile Technology Digest
World Textile Abstracts

VETERINARY SCIENCE
*Index Veterinarius
*Veterinary Bulletin

WILDLIFE AND FISHERIES
*Aquatic Science and Fisheries Abstracts
Commerical Fisheries Abstracts
Fisheries Review
Marine Fisheries Abstracts
Ocean Abstracts
Sport Fishery Abstracts
Wildlife Abstracts
Wildlife Research
Wildlife Reviews
World Fisheries Abstracts

✦ EXERCISES

1. Using an index appropriate for your field, look up a current issue you may have found in one of your science textbooks. Find the title of one current article in a scholarly journal as listed by the index. Write down or print out the citation from the index, find the complete journal name in the front of the index, and using that information find the journal article in your library. (Remember, you will need to look up the journal name in the serials listing for your particular library.)

   Browse the nearby shelves to see which other journals from that field are carried by your library. Make a photocopy of either the article you located in the index or another interesting article that you ran across in your search. Read the article carefully; on your photocopy, underline key ideas. Then construct an outline of the article. Turn in the photocopy with your outline.

2. Find any information your library has about online computer searches and discuss computer searching in your particular field with the librarian responsible for computer searches.

3. Find out whether your library has any databases on videodisk or CD-ROM that you can search using a microcomputer. Try out any such tools, using appropriate subject headings and key words for your topic, and turn in the printout generated from your search.

4. Interview one of your professors about library research in his or her field. How does that professor gather information needed for his or her work? How has the research process changed over the years? How much does the professor rely on computer searching? Write a short paragraph summarizing the interview.

5. Interview a reference librarian at your campus library. How has information access changed? How many new information sources have come to campus in the past year or so? What does the librarian foresee in the libraries of the future? Write a short paragraph summarizing the interview.

3

Library Research Methods

Successful research depends on knowing what library resources are available, but it also depends on knowing how to find and use those resources. Developing a search strategy will help you to find materials on your research topic and to use library resources efficiently. First, you need to use library tools to locate source materials; then you must evaluate those sources and interpret them so that they will be useful to your particular research project.

PREPARATION AND INCUBATION

As a college student, you are uniquely prepared to conduct a research project. Your experiences and prior schooling have given you a wealth of information to draw from. A research project may begin in one of the following ways: you may have been assigned to do a research paper in a particular college class or you may have discovered an interesting question or problem on your own that you decided to investigate. Although the former impetus, a course assignment, might seem artificial or contrived at first, in reality it may give you the opportunity to investigate something that has always intrigued you.

Finding a Topic

A good place to begin looking for a research topic is in the textbooks you are currently using for the courses you are taking. Scan the table of contents with an eye toward a topic that you'd enjoy investigating. Or if you have no idea at all, begin by browsing through a specialized encyclopedia, such as the *Encyclopedia of Bioethics* or the *Encyclopedia of Physics*. A third resource for finding topics is Editorial Research Reports (ERR). If your library subscribes to ERR, it receives a weekly description of a wide variety of contemporary events, problems, and issues, such as acid rain, homelessness, or women in business. A bibliography listing several recent articles, books, and reports on each topic covered by ERR is also included. Skimming these reports may spark an interest in a particular topic, as well as get you started on finding relevant materials and information.

Once you have an idea for a topic, you might discuss your idea with a reference librarian, with your teacher, and with other students in your class. They may have ideas or suggestions related to your topic or may be able to direct you to aspects of the topic you may not have considered. Your goal is to focus your topic by asking pertinent starting questions that your research will attempt to answer. So, if your general topic is "acid rain," your specific starting question might be "What have the effects of acid rain been on the forests of New York?" Or, "What is the EPA currently doing to control pollutants that cause acid rain?" Such questions, which should be neither too trivial nor too broad, will help you to sort through information in search of an answer and may prevent you from aimlessly reading on a topic area that is too general.

Gathering Research Materials

The Research Notebook

To keep track of all your research, you need to obtain a notebook to serve as your "research notebook." In the notebook, you record your specific topic area and the starting questions you wish to answer, outline your plans for searching the library (called a library search strategy), and begin a list of sources (called a working bibliography). Your research notebook is also the place where you can begin to articulate for yourself your own understanding of the answer to your starting question as it evolves through your research. It is crucial that, as you investigate your topic, you record in the research notebook not only what others have said on the subject, but also your own impressions and comments.

Your research notebook, then, will be the place for tracing your entire research process: the starting questions, the search strategy, the sources used in your search, your notes, reactions to and comments on the sources, the tentative answers you propose to your starting question, a thesis statement articulating the main points to be covered by your paper, an informal outline or an organizational plan for your paper, and all preliminary drafts of your paper. Many students like to take notes from sources in their research notebooks instead of on notecards, in order to keep all their research information in a single convenient place. If you decide on this approach, be sure to record your notes and evaluative comments in separate parts of the notebook. I suggest to my students that they leave a blank page for comments in their notebooks adjacent to each page of notes. Also, remember to reserve a place in your notebook for recording any primary research data (such as interview, survey, or questionnaire data) that you collect in connection with your research project.

A Computer Research Notebook

If you are using a word processor, you can take advantage of the storage capabilities of your computer to develop a computer research notebook. Create a file or directory on your computer and label it your research notebook file. All of the items previously listed for a research notebook can be gathered together in this one file or directory on your computer. For example, you could include your topic and starting questions, search strategy, working bibliography, notes and evaluative comments on sources, and so on. In this way, your research project can proceed systematically as you gather information and build your own expertise on your topic through your evolving computer file. Using word processing, then, you can revise your notebook file, organize your information, and even write your paper based on the stored information on disk.

For example, a student who had to write a research report for a computer science class decided to write about computer crime. In thinking about the subject, he determined that first he needed to categorize the types of computer crimes so that he could arrange the information in his research notebook in an organized way. As he read books and articles on the subject, he began to sort materials into the following subheadings: computer as object or target of crime, computer as subject or site of crime, computer as instrument used to create crime, and computer as symbol for criminal deception or intimidation. After entering these subheadings onto his computer file, he gradually built up the report; as he encountered information for the various sections of his report, he added it under the appropriate subheading in his file. You could use a similar method for your own research notebook.

The Working Bibliography

A bibliography, as you learned in Chapter 2, is a list of books and articles on a particular subject. Your working bibliography, your preliminary list of sources, grows as your research progresses, as one source leads you to another. It is called a "working" bibliography—as opposed to the finished bibliography—because it may contain some sources that you ultimately will not use in your paper.

A working bibliography need not be in final bibliographic form, but it is important to record accurately all the information eventually needed to compose your final bibliography to keep from having to backtrack and find a book or article again. Oftentimes, a student finds a book, reads a relevant section and takes notes, but neglects to write down all of the bibliographic information, that is, the author(s), complete title, publisher, date and place of publication, and so on (see p. 19). Then, when compiling the final bibliography in which are listed all of the sources referred to in the paper, the student discovers that he or she has not written down the date of publication, for instance, or the author's first name. This means another trip to the library to find the book or journal, which may or may not still be on the shelf!

The working bibliography should be comprehensive, the place where you note down all sources that you run across—in bibliographies or databases, for example, whether your library has them and whether they turn out to be relevant to your topic. So, the working bibliography is a complete record of every possible path you encountered in your search, whether or not you ultimately followed that path. In contrast, the final bibliography lists only those sources that you actually read and used as references for your own paper.

The example below of a working bibliography comes from a student's research project on the body's immune system. The student used a citation style commonly used in sciences and technology.

```
        Working Bibliography
Golub, E. S. 1987. Immunology: A synthesis.
     Boston: Sinauer Associates.
Bellanti, J. A. 1978. Immunology II.
     Philadelphia: W. B. Saunders.
Bass, A. 1985. Unlocking the secrets of immunity.
     Technology Review 88: 62.
Getzoff, E. 1987. Mechanisms of antibody binding
     to a protein. Science 235: 1191.
Herscowitz, H. 1985. Cell-mediated immune
     reactions. Philadelphia: W. B. Saunders.
```

Silberner, J. 1986. Second T-cell receptor found. Science News, 180: 50.

Wise, H. 1987. Man bites man. Hippocrite 100: 93.

Quinn, L. Y. 1968. Immunological concepts. Ames, IA: Iowa State University Press.

Reissig, J. L. 1977. Microbial interactions. New York: Chapman and Hall Press.

Developing a Search Strategy

Once you have decided on some starting questions and have gathered the necessary research materials, you are ready to outline a preliminary search strategy. Many library research projects begin in the reference area of the library, since the library tools that refer you to other sources are kept there. Often you will begin with reference works (dictionaries, encyclopedias, and biographies), in which can be found background and contextualizing information on your topic. Then, you may proceed to more specific reference works (abstracts, indexes, and databases). To make your library search an orderly and thorough process, you should design a search strategy, beginning with general sources and working to more specific sources. In most fields, a search strategy includes the following major components:

1. Background sources—dictionaries and encyclopedias (including discipline-specific sources)
2. Reviews of literature and research reports (to discover how others outline or overview the subject)
3. Print indexes and bibliographies (for listings of source materials by subject) (Remember to use the LCSH for subject headings)
4. Library online catalog and other databases (including CD-ROMs) for subject and keyword searching of books and journal articles on your topic (Use your library's serials listing to locate sources available in your library)
5. Primary research (for firsthand information such as interviews or surveys)

One student designed the following search strategy to help him begin his research on computer crime:

1. Look up "computer crime" in dictionaries and encyclopedias, including *Chambers' Dictionary of Science and Technology* and *The Encyclopedia of Computer Science and Engineering* for definitions.

2. Look up any reviews already done on computer crime, using the *Index to Scientific Reviews* and *Current Contents* for background overviews on the subject.
3. Use the *Applied Science and Technology Index* to look up current works on computer crime. Check headings to be sure that the keyword is "computer crime." (Note: This student discovered that this particular index listed articles on computer crime under the heading "Electronic Data Processing—Security Measures.") Use the *Science Citation Index* for a forward search on key sources.
4. Look up "computer crime" in the *Library of Congress Subject Headings* (LCSH) list to determine whether it is the subject heading used in the online catalog. Then, using the online catalog, search for sources on "computer crime" and other related headings by subject and keywords to find books and articles on computer crime.

Notice that this student's search began in the reference area and ended with use of the online catalog. Many students make the mistake of using the catalog before they really know enough about their topics to make it useful. You would be wise to order your search strategy to begin in the reference area as well.

You may need to change or modify your search strategy as you go along; do not feel that the strategy must be rigid or inflexible. However, using a search strategy enables you to proceed in an orderly, systematic fashion with your research. On your working bibliography, write down the complete citation for each source you encounter in your search. As you read in the general and specialized encyclopedias, for example, you may find related references listed at the end of articles. Write down in your working bibliography complete citations for any references that look promising so that you can look them up later. Similarly, as you look through the reviews, the indexes, and the card catalog, write down the citations to any promising sources. In this way, you will build your working bibliography during your library search.

✦ EXERCISES

To begin preparation for your own research project, follow these steps:

1. Select and narrow a research topic, that is, limit the topic in scope so that it is of a manageable size. Talk your topic over with others, including your classmates, your teacher, and your librarians.

2. Articulate several starting questions that you would seek to answer during your research.
3. Gather your research materials—notebook and notecards. Or create a computer file or directory for a computerized notebook.
4. Reserve space in your research notebook or computer file for both notes from sources and the evaluative comments that you will write down as you are reading.
5. Reserve space in your research notebook or computer file for your working bibliography in which you will list all of the sources you encounter during your search.
6. Outline your search strategy (refer to Chapter 2 for specific library tools to use in your search and to Chapters 6, 7, 8, or 9 for a sample search in your chosen discipline).

Outlining a Time Frame

After writing down your search strategy, you will have at least some idea of how long your research is likely to take. Now is the time to sit down with a calendar and create a time frame for your entire research project. Your teacher may have given you some deadlines, and if so, they will help you decide on a time frame. If not, you will have to set your own dates for accomplishing specific tasks so that you can proceed in an orderly fashion toward the completion of the project. If you have never done a research project before, you might be overwhelmed at the thought of such a large task. However, if you break the job down into smaller parts, it will seem more manageable.

Allow yourself three to four weeks for locating, reading, and evaluating sources. As you begin to work in the library, you will see that a library search is a very time-consuming process. Just locating sources in a large library takes time; perhaps one book you need will be shelved in the third sub-basement and another on the fourth floor! Sometimes a book you want will have been checked out; in such a case you will have to submit a "recall notice" to the librarian asking that the book be returned and reserved for you. You may also find that you need to obtain materials from another library through an interlibrary loan, another time-consuming process. Plan to spend two to three hours in the library each day for the first month of your research project. After that, you may find you can spend less time in the library.

If your research project involves primary research (see Chapter 4), begin to plan for that research while you are writing your search strategy. Allow one to two weeks for conducting your primary research, depending on its nature and scope.

Schedule one to two weeks for preliminary writing. To make sense of your subject and answer your starting question, you need to spend time and effort in studying and evaluating your sources, in brainstorming and writing discovery drafts. Eventually, you ought to be able to express your understanding of the subject in a thesis statement, which helps control the shape and direction of the research paper and provides your readers with a handle on your paper's main idea or argument.

Finally, give yourself enough time to plan, organize, and write the complete draft of your research paper. You need time to plan or outline your paper and to construct your argument, using your source information to reinforce or substantiate your findings in a clearly documented way. Allow yourself one to two weeks for organizing and writing rough drafts of your paper and an additional week for revising, polishing, and editing your final draft. If you intend to hire a typist or if you need to type your paper in what may turn out to be a busy computer lab, allow an extra week for the typing of the paper.

As you can see from this overview, most research projects take an entire college term to complete. Recall from Chapter 1 the stages in the research process: preparation, incubation, illumination, and verification. You need to consider all four stages as you plan your research project. Allow time for your library search, time for ideas to incubate in your subconscious, time for arriving at an understanding of your topic, and time to verify that understanding in writing.

What follows is a sample time frame to give you some idea of how you might budget your own time:

Week 1: Select preliminary research topic; articulate starting questions; gather and organize research notebook; draw up tentative search strategy; plan research time frame; read general background sources; begin to focus topic.

Week 2: Build working bibliography by using indexes, online catalogs, and databases; begin to locate sources in library.

Week 3: Read and evaluate sources; take notes on relevant sources; in research notebook, comment on sources, that is, their importance to topic and their relationship to other sources.

Week 4: Arrange and conduct any primary research; complete reading and evaluating of sources; identify gaps in research and find more sources if necessary.

Week 5: Begin preliminary writing in research notebook— summary, synthesis, critique activities; initiate brainstorming and discovery drafting; begin to define an answer to the starting question.

Week 6: Write a tentative thesis statement; sketch a tentative plan or outline of the research paper.

Week 7: Write a rough draft of the research paper; keep careful track of sources through accurate citation (distinguish quotes from paraphrases).
Week 8: Revise and edit the rough draft; spellcheck; check correct usage and documentation of sources.
Week 9: Print and proofread final copy carefully; have a friend or classmate proof as well.

✦ EXERCISES

1. Outline the time frame of your research project; refer to a current academic calendar from your school and to any deadlines provided by your teacher.

2. Plan any primary research you intend to conduct for your project. For example, if you need to contact someone for an interview, do so well ahead of time.

LOCATING SOURCES

After defining your search strategy and outlining a time frame for your project, you can begin the actual research process in the library. Refer to the relevant section of Chapter 2, "Library Resources."

✦ EXERCISE

Begin your library search, writing down source citations in your working bibliography. Refer to the relevant section of Chapter 2 and to the model search outlined above for explicit direction in the process of research for your chosen topic.

WORKING WITH SOURCES

One of the most crucial aspects of the research process is the development of the skills needed to pull the appropriate information from the source materials you have gathered.

Reading for Meaning

The sources you locate in your library search are the raw material for your research paper. You might supplement these sources with primary data, but generally your research paper will be based on information from written secondary sources. Your job is to read carefully and actively. Reading is not a passive process by which the words float into your mind and become registered in your memory. If you read passively, you will not comprehend the author's message. You have probably had the experience of rereading a passage several times and still not understanding a word of it. In such cases, you were not reading actively. In active reading, the reader is engaged in a dialogue with the author.

To be fair in your interpretation of what you are reading, you must first be receptive to what the author is saying; approach anything you read with an open mind. Before actually beginning to read, look at any nontextual materials that accompany the source, including information about the author, the publisher, the origin of the work, the title, and the organizational plan or format of the work.

The Author. Questions to consider about the author include the following: Who is the author? Is the author living or dead? What other works did the author write? What are the author's qualifications and biases on this particular topic? Is the author affiliated with any organizations that might espouse a particular point of view (e.g., the National Rifle Association or the National Organization for Women)? Is the author a faculty member at a reputable college or university? Does the author work for a government agency, a political group, or an industry?

The Publication. You should look at the publication information for the source. What was the date of publication? Would the particular context help you understand the work? It is also important to evaluate the medium in which the work appears. Was this an article published in a popular magazine, such as *Ladies' Home Journal*, or was it a paper in a scholarly journal, such as *College and Research Libraries*? The title of the work itself may reveal some important information. Does the title seem significant? Does it indicate the probable conclusions or main points of the article or book? Does it provide a clue to some controversy? Authors provide titles as indicators of what the work will contain, so we should be sensitive to the title of any work we read.

The Organization. Finally, before you read the work, look at the author's organizational plan. For a book, look at the table of contents, the introduction or preface, the chapter headings, and the major subdivisions. Judging from these clues, ask yourself what the author's main points are likely to be. What does the author seem to consider important about the subject? For articles, read the abstract (if provided) and look at any subdivisions or headings. These will help you to understand the overall structure of the article.

Active Reading and Notetaking

Once you have completed your preliminary overview of the work, you are ready to begin the actual reading process. Plan to read a work that you need to understand thoroughly at least twice. During the first reading, go through the work at a relatively quick pace, either a section or a chapter at a time. If an article is relatively short, you can read it through entirely at one sitting. As you go along, pay attention to key words or phrases and try to get a general idea of the author's main points. This reading will be more than a skimming of the work; you should be able to generally understand what you have read on this first time through.

Then read the work again, carefully and slowly. Use a highlighter or a pencil to underline key ideas and to write in the margins of your own books or photocopied articles, called annotating the article. Of course, if you are borrowing a book from the library, you will not be able to underline on the book itself. In that case, record key ideas, using your own words, on your notecards or in your research notebook, along with the author's name and the page number on which the material was found.

The second time you read, stop frequently to absorb the information and interpret it in your own mind. For each paragraph or section of the article or chapter, jot down on the page itself or on notecards a summary sentence or two that captures the main idea. Be sure that these marginal notes are in your own words. It is crucial that you paraphrase the author's ideas and that you note down the page number on which you found the information.

Paraphrasing Appropriately and Avoiding Plagiarism

Paraphrasing. Paraphrasing may be defined as restating or rewording a passage from a text, giving the same meaning in another form. The objective of paraphrasing, then, is to present an author's ideas in your own words. When paraphrasing fails, it may be because the reader misunderstood the passage, the reader insisted on reading his or her own ideas into the passage, or the reader partially understood but chose to guess at the meaning rather than fully understanding it. To paraphrase accurately, you must first read closely and understand completely what you are reading. Here are five suggestions that will help you as you paraphrase:

1. Place the information found in the source in a new order.
2. Break the complex ideas into smaller units of meaning.
3. Use concrete, direct vocabulary in place of technical jargon found in the original source.
4. Vary the sentence patterns.
5. Use synonyms for the words in the source.

Plagiarism. Plagiarism is defined as the unauthorized use of the language and thoughts of an author and the representation of them as one's own. Oftentimes students taking notes from a source will inadvertently commit plagiarism by careless copying of words and phrases from the author that end up in the student's paper and appear as if written by the student himself or herself. If you set the original piece aside while you are taking notes, you are less likely to copy the author's wording. Then go back and double check your notes against the original.

Sometimes plagiarism is not inadvertent at all, but rather overt theft of one author's work by another, as the accompanying article by Gregg Easterbrook attests. However, whether the plagiarist is a Stanford author or a newspaper reporter, it is always unethical to present someone's words or ideas as if they were your own. Nor is providing a footnote alone always sufficient. Any of the author's words or phrases should be enclosed in quotation marks to signal to the reader that exact wording and phrasing as written by the original author is being used; in addition, a footnote or parenthetical citation to the source should be provided.

The Sincerest Flattery

Thanks, but I'd rather you not plagiarize my work

Gregg Easterbrook

It was the best of times, and pretty much the worst of times. I felt borne back ceaselessly to the past. Maybe that's because days on the calendar creep along in a petty pace, and all our yesterdays but light fools the road to dusty death.

OK, the above words are not really mine. But hey, I changed them slightly. I thought nobody would notice.

Some kind of harmonic convergence of plagiarism seems to be in process. A Boston University dean, H. Joachim Maitre, was caught swiping much of a commencement address from an essay by the film critic Michael Medved. Fox Butterfield of *The New York Times* then cribbed from a *Boston Globe* story about the swipe. The *Globe*, in turn, admitted that one of its reporters was disciplined for stealing words from the Georgia politician Julian Bond. Laura Parker of *The Washington Post* (which owns NEWSWEEK) was found poaching from The *Miami Herald*'s John Donnelly. And the president of Japan's largest news service, Shinji Sakai, announced his resignation, taking public responsibility for 51 plagiarized articles that were discovered last May.

I personally entered the arena last week when the *Post* reported that a Stanford University business-school lecturer plagiarized me in a recent book, "Managing on the Edge." Chapters about the Ford Motor Co. contain approximately three pages nearly identical in wording to an article on Ford that I wrote five years ago for The Washington Monthly magazine.

What's it like to discover someone has stolen your words? My initial reaction was to feel strangely flattered that another author had liked my writing well enough to pass it off as his own.

OK, I plagiarized that last sentence. It comes from the writer James Fallows, who last week in a National Public Radio commentary described the two times he has been plagiarized. In one case a San Jose State professor published a textbook in which an entire chapter was nearly identical to an article Fallows had written. For good measure, another chapter was nearly identical to an article by the economic analyst Robert Reich. The professor claims this happened because of computer error. The publisher sent Fallows a letter saying that, since we disseminated your copyrighted work without permission, could we have permission now?

Plagiarism is the world's dumbest crime. If you are caught there is absolutely nothing you can say in your own defense. (Computer error?) And it's easy to commit the underlying sin—presenting as your own someone else's work—without running the risk of sanction, merely by making the effort to reword.

Yet figures as distinguished as Alex Haley, John Hersey, Martin Luther King Jr., and D. M. Thomas have been charged with borrowing excessively from the work of others. One factor is sloth. Another is ego: there are writers who cannot bear even tacit admission that someone else has said something better than they could. The line between being influenced by what others have written, and cribbing from it outright, is one nearly every writer walks up to at some point.

About that last sentence. The *Post* quoted me as saying that: does that mean I just plagiarized myself? It can happen. Conor Cruise O'Brien was accused of self-plagiarizing when he sold, to *The Atlantic,* an article hauntingly similar to one previously run under his byline in *Harper's*.

Perhaps word rustlers tell themselves they will never be caught, and indeed, unlikely combinations of events may be necessary for a theft to be exposed. Some no doubt further tell themselves that if they are caught no one will sue. Most writers don't make serious money and so are uninviting targets for litigation.

I might never have learned about "Managing on the Edge" if an alert reader named Robert Levering had not been researching Ford Motor Co. Shortly after reading my article, he saw a "Managing" excerpt from the Stanford business-school magazine. He not only realized he was reading the same words, but more important, remembered where he encountered them first. Levering wrote to Stanford. I heard about his letter, got a copy of the book, and my jaw dropped. Particularly galling, on the book's facing page, was the phrase "Copyright 1990 by Richard Tanner Pascale." The author was asserting ownerships of words I composed.

Elegantly crafted: Unlike the Boston University incident, where the dean stole words for an unpaid speech, in this case there was money involved. "Managing" is a commercial book published by Simon & Schuster, a reasonable seller with 35,000 copies in print and another run pending. It's been well received by critics: mainly, I suspect, because of three particularly elegantly crafted pages. Another person was not only presenting my words as his own, but doing so for gain.

The author of "Managing" apologized for what happened but contended it was not plagiarism, because my name is in the book's footnotes. Footnotes my foot. Footnotes mean the place a fact can be found; they do not confer the right to present someone else's words as your own work. Any Stanford undergraduate who attempted that defense would not get far.

My case dragged on inconclusively for a while. But the moment a reporter for *The Washington Post* called Simon & Schuster, the pace of cooperation accelerated dramatically. Simon & Schuster is now preparing corrections for future editions of "Managing." Stanford has an academic committee investigating its end of the incident.

The wave of plagiarism disclosures poses an obvious question: are dozens of authors now quaking in their shoes, worried about whether some alert reader will stumble across the resemblance between those pages in their book and, say, that article in some obscure little journal no one ever reads?

Frankly, Scarlet, I don't give a damn. ✦

Easterbrook, G. (1991, July 29). The sincerest flattery: Thanks, but I'd rather you not plagiarize my work. *Newsweek*, pp. 45-46.

How to Paraphrase Appropriately

The examples below show acceptable and unacceptable paraphrasing:

ORIGINAL PASSAGE

During the last two years of my medical course and the period which I spent in the hospitals as house physician, I found time, by means of serious encroachment on my night's rest, to bring to completion a work on the history of scientific research into the thought world of St. Paul, to revise and enlarge the *Question of the Historical Jesus* for the second edition, and together with Widor to prepare an edition of Bach's preludes and fugues for the organ, giving with each piece directions for its rendering. (Albert Schweitzer, *Out of My Life and Thought*. New York: Mentor, 1963, p. 94.)

A POOR PARAPHRASE

Schweitzer said that during the last two years of his medical course and the period he spent in the hospitals as house physician he found time, by encroaching on his night's rest, to bring to completion several works.

Embarrassing Echoes

In both journalism and academia, plagiarism is close to mortal sin. Sometimes writers are tempted to stray, giving themselves credit for the work of another. A side-by-side comparison can be withering.

Gregg Easterbrook, Oct. 1986:
"On a very dark day in 1980, Donald Peterson, newly chosen president of Ford Motors, visited the company design studios. Ford was in the process of losing $2.2 billion, the largest single-year corporate loss in U.S. history."

Richard Pascale, March, 1990
"On a dark day in 1980, Donald Peterson, the newly chosen President of Ford Motor Company, visited the company's Detroit design studio. That year, Ford would lose $2.2 billion, the largest loss in a single year in U.S. corporate history."

Sources: The Washington Monthly: "Managing on the Edge"

Michael Medved, Feb. 2, 1991:
"Apparently, some stern decree has gone out from the upper reaches of the Hollywood establishment that love between married people must never be portrayed on the screen."

H. Joachim Maitre, May 12, 1991:
"Apparently, some stern advice has come from the upper reaches of the Hollywood establishment that love between married people must never be portrayed on the screen."

Source: The Boston Globe

[Note: This paraphrase uses too many words and phrases directly from the original without putting them in quotation marks and thus is considered plagiarism. Furthermore, many of the ideas of the author have been left out, making the paraphrase incomplete. Finally, the student has neglected to acknowledge the source through a parenthetical citation.]

A GOOD PARAPHRASE

```
    Albert Schweitzer observed that by staying
up late at night, first as a medical student and
then as a "house physician," he was able to
finish several major works, including a
historical book on the intellectual world of St.
Paul, a revised and expanded second edition of
Question of the Historical Jesus, and a new
edition of Bach's organ preludes and fugues
complete with interpretive notes, written
collaboratively with Widor (Schweitzer 94).
```

[Note: This paraphrase is very complete and appropriate; it does not use the author's own words, except in one instance, which is acknowledged by quotation marks. The student has included a parenthetical citation that indicates to the reader the paraphrase was taken from page 94 of the work by Schweitzer. The reader can find complete information on the work by turning to the bibliography at the end of the student's paper.]

Making Section-by-Section Summaries

As an alternative to close paraphrasing, you may wish to write brief summaries (three or four sentences) on your notecards or in your notebook. Again, use your own words when writing these summaries. If the material is particularly difficult, you may need to stop and summarize more frequently than after each section or chapter. If it is relatively simple to understand or not particularly pertinent to your topic, take fewer notes and write shorter summaries. At any rate, be certain that you are internalizing what you read—the best gauge of your understanding of the material is your ability to put it into your own words in the form of paraphrases or short section-by-section summaries. Again, as with paraphrases or quotes, note down the page numbers on which the material was found.

Reviewing

After completing your marginal notes, paraphrases, or summaries, go back and review the entire piece, taking time to think about what you read. Evaluate the significance of what you learned by relating the work to your own project and starting questions. Your research notebook is the place to record the observations and insights gained in your reading. How does the work fit in with other works you read on the same topic? What ideas seem particularly relevant to your own research? Does the work help to answer your starting question? Answering such questions in your research notebook will help you to put each work you read into the context of your own research.

Perceiving the Author's Organizational Plan

In writing, you should attempt to make your organizational plan clear to your potential readers. Similarly, while reading, you should attempt to discern the organizational plan of the author. One of the best ways to understand the author's plan is to try to reconstruct it through outlining. For an article or book that seems especially important to your research project, you may want to understand the material in a more complete and orderly way than that gained through paraphrasing or summarizing. You can accomplish this goal by constructing an outline of what you have read.

In a well-written piece, the writer will have given you clues to important or key information. Your summaries should have identified main ideas that are most likely to be the main points of the outline. However, you may still need to go back to the work to identify the author's secondary, supporting points, including examples, illustrations, and supporting arguments used to make each individual argument clearer or more persuasive. In outlining a key source, you can come to understand it more fully. Again, be sure that all the points in your outline have been stated in your own words rather than the words of the author.

♦ **EXERCISE**

The following article has been included to illustrate how to go about underlining, annotating, summarizing, and outlining a key source. Read the article carefully, noticing which ideas have been underlined and which annotated. Do you agree with my identification of key ideas? Why or why not? Is there a right or wrong identification of key ideas? The final third of the article has not been underlined, summarized, or included on the outline that follows. Try out these four techniques (underlining, annotating, summarizing, outlining) by finishing the interpretation of the article, beginning immediately after the quotation from Paul Dirac:

1. Underline key ideas.
2. Annotate the article by putting notes in the margins that paraphrase the author's words.
3. Summarize in your own words the main ideas of the last section of the article, as you would on a notecard or in your research notebook (three to four sentences).
4. Complete the outline following the article, using your own marginal notes and annotations on the article.

The Scientific Aesthetic

K. C. *Cole*

Opening quote from physics textbook.

"Poets say science takes away from the beauty of the stars—mere globs of gas atoms. Nothing is 'mere.' I too can see the stars on a desert night, and feel them. But do I see less or more? The vastness of the heavens stretches my imagination—stuck on this carrousel, my little eye can catch one-million-year-old light . . . For far more marvelous is the truth than any artists of the past imagined! Why do the poets of the present not speak of it? What men are poets who can speak of Jupiter if he were like a man, but if he is an immense spinning sphere of methane and ammonia must be silent?"

Feynman rejects idea that science makes nature ugly.

There is both beauty and creativity in science and art, says Feynman.

This poetic paragraph appears as a footnote in, of all places, a physics textbook: *The Feynman Lectures on Physics* by Nobel laureate Richard Feynman. Like so many others of his kind, Feynman scorns the suggestion that science strips nature of her beauty, leaving only a naked set of equations. Knowledge of nature, he thinks, deepens the awe, enhances the appreciation. But Feynman has also been known to remark that the only quality art and theoretical physics have in common is the joyful anticipation that artists and physicists alike feel when they contemplate a blank piece of paper.

What is the kinship between these seemingly dissimilar species, science and art? Obviously there is some—if only

Working with Sources 51

What is the relationship between science and art? There must be a link, as so many scientists are also artists.

because so often the same people are attracted to both. The image of Einstein playing his violin is only too familiar, or Leonardo with his inventions. It is a standing joke in some circles that all it takes to make a string quartet is four mathematicians sitting in the same room. Even Feynman plays the bongo drums. (He finds it curious that while he is almost always identified as the physicist who plays the bongo drums, the few times that he has been asked to play the drums, "the introducer never seems to find it necessary to mention that I also do theoretical physics.")

Art and science cover same ground—examples.

<u>One commonality is that art and science often cover the same territory. A tree is fertile ground for both the poet and the botanist.</u> The relationship between mother and child, the symmetry of snowflakes, the effects of light and color, and the structure of the human form are studied equally by painters and psychologists, sculptors and physicians. The origins of the universe, the nature of life, and the meaning of death are the subjects of physicists, philosophers, and composers.

Differing approaches: art = emotion, science = logic; but many scientists disagree, arguing that emotion in science is integral to the process.

<u>Yet when it comes to approach, the affinity breaks down completely. Artists approach nature with feeling; scientists rely on logic. Art elicits emotion; science makes sense.</u> Artists are supposed to care; scientists are supposed to think.

At least one physicist I know rejects this distinction out of hand: "What a strange misconception has been taught to people," he says. "They have been taught that one cannot be disciplined enough to discover the truth unless one is indifferent to it. Actually, there is no point in looking for the truth unless what it is makes a difference."

The history of science bears him out. Darwin, while sorting out the clues he had gathered in the Galapagos Islands that eventually led to his theory of evolution, was hardly detached. "I am like a gambler and love a wild experiment," he wrote. "I am horribly afraid." "I trust to a sort of instinct and God knows can seldom give any reason for my remarks." "All nature is perverse and will not do as I wish it. I wish I had my old barnacles to work at, and nothing new."

Examples of scientific passion—Einstein.

The scientists who took various sides in the early days of the quantum debate were scarcely less passionate. Einstein said that if classical notions of cause and effect had to be renounced, he would rather be a cobbler or even work in a gambling casino than be a physicist. Niels Bohr called Einstein's attitude appalling, and accused him of high treason.

Another major physicist, Erwin Schrodinger, said, "If one has to stick to this damned quantum jumping, then I regret having ever been involved in this thing." On a more positive note, Einstein spoke about the universe as a "great, eternal riddle" that "beckoned like a liberation." As the late Harvard professor George Sarton wrote in the preface to his *History of Science*, "<u>There are blood and tears in geometry as well as in art</u>."

Instinctively, however, most people do not like the idea that scientists can be <u>passionate</u> about their work, any more than they like the idea that poets can be calculating. <u>But it would be a sloppy artist indeed who worked without tight creative control, and no scientist ever got very far by sticking exclusively to the scientific method</u>. Deduction only takes you to the next step in a straight line of thought, which in science is often a dead end. "Each time we get into this log jam," says Feynman, "it is because the methods we are using are just like the ones we have used before . . . A new idea is extremely difficult to think of. It takes fantastic imagination." <u>The direction of the next great leap is as often as not guided by the scientist's vision of beauty</u>. Einstein's highest praise for a theory was not that it was good but that it was beautiful. His strongest criticism was "Oh, how ugly!" He often spoke about the aesthetic appeal of ideas. "Pure logic could never lead us to anything but tautologies," wrote the French physicist Jules Henri Poincaré. "It could create nothing new; not from it alone can any science issue."

Poincaré also described the role that aesthetics plays in science as "a delicate sieve," an arbiter between the telling and the misleading, the signals and the distractions. Science is not a book of lists. The facts need to be woven into theories like tapestries out of so many tenuous threads. Who knows when (and how) the right connections have been made? <u>Sometimes, the most useful standard is aesthetic</u>: Erwin Schrodinger refrained from publishing the first version of his now famous wave equations because they did not fit the then-known facts. "I think there is a moral to this story," Paul Dirac commented later. "Namely, that it is more important to have beauty in one's equations than to have them fit experiment . . . It seems that if one is working from the point of view of getting beauty in one's equations, and if one has really a sound insight, one is on a sure line of progress."

Sometimes the connection between art and science can be even more direct. Danish physicist Niels Bohr was known

---

Sidenotes:

Artists and scientists are both passionate and in control of their work.

Creativity needed in science; control needed in art.

Scientists often proceed based on their own "vision of beauty."

Illustrations from science: aesthetics serve as "delicate sieve" for science.

for his fascination with cubism—especially "that an object could be several things, could change, could be seen as a face, a limb, and a fruit bowl." He went on to develop his philosophy of complementarity, which showed how an electron could change, could be seen either as a particle or a wave. Like cubism, complementarity allowed contradictory views to coexist in the same natural frame.

Some people wonder how art and science ever got so far separated in the first place. The definitions of both disciplines have narrowed considerably since the days when science was natural philosophy, and art included the work of artisans of the kind who build today's fantastic particle accelerators. "Science acquired its present limited meaning barely before the nineteenth century," writes Sir Geoffrey Vickers in Judith Wechsler's collection of essays *On Aesthetics in Science*. "It came to apply to a method of testing hypotheses about the natural world by observations or experiments. . . ." Surely, this has little to do with art. But Vickers suspects the difference is deeper. People want to believe that science is a rational process, that it is describable. Intuition is not describable, and should therefore be relegated to a place outside the realm of science. "Because our culture has somehow generated the unsupported and improbable belief that everything real must be fully describable, it is unwilling to acknowledge the existence of intuition."

There are, of course, substantial differences between art and science. Science is written in the universal language of mathematics; it is, far more than art, a shared perception of the world. Scientific insights can be tested by the good old scientific method. And scientists have to try to be dispassionate about the conduct of their work—at least enough so that their passions do not disrupt the outcome of experiments. Of course, sometimes they do: "Great thinkers are never passive before the facts," says Stephen Jay Gould. "They have hopes and hunches, and they try hard to construct the world in their light. Hence, great thinkers also make great errors."

But in the end, the connections between art and science may be closer than we think, and they may be rooted most of all in a person's motivations to do art, or science, in the first place. MIT metallurgist Cyril Stanley Smith became interested in the history of his field and was surprised to find that the earliest knowledge about metals and their properties was provided by objects in art museums. "Slowly, I came to see that this was not a coincidence but a consequence of the very nature of discovery, for discovery derives from aesthetically motivated curiosity and is rarely a result of practical purposefulness." ✦

Cole, K. C. (1983). *The Scientific Aesthetic*. New York: Discover Publications, Inc. Reprinted by permission.

OUTLINE BASED ON MARGINAL NOTES

I. Opening quote from physics text to introduce the topic (Feynman)
   A. Feynman rejects the idea that science makes nature ugly
   B. There is both beauty and creativity in science and art, says Feynman
II. What is the relationship between science and art?
   A. There must be a link because so many scientists are also artists
   B. Art and science cover the same ground (examples to illustrate)
   C. Differing approaches: art is emotional, science is logical
   D. But many scientists disagree; they argue that emotion, caring in science is integral to the process.
   E. Examples of scientific passion, including Einstein
III. The importance of control and creativity in both science and art
   A. Scientists often proceed based on their own "vision of beauty"
   B. Illustrations from science
   C. Aesthetics serves as "a delicate sieve" for science
IV. Links between art and science can be quite straightforward

[Continue outline on a separate sheet of paper.]

ILLUMINATION AND VERIFICATION

An essential part of your research is the evolution of your understanding of the subject. As you read and evaluate your sources, you will be seeking a solution to your starting question. Several preliminary writing tasks can help you evaluate your sources and understand your topic better.

Evaluation

In your working bibliography, you record the information needed to find a source in the library. Once you have located a source, you need to evaluate it for its usefulness to your particular research project and to your starting question. Every library search will entail the systematic interaction of examination, evaluation, and possibly elimination of material. It is not unusual for an article with a promising title to turn out to be totally irrelevant. Do not be discouraged by dead ends of this sort—they are an accepted and expected part of the library search. You must not hesitate to eliminate irrelevant or unimportant information. As you read each source, consider the following criteria (see also the section on Active Reading).

Evaluative Criteria

1. The relevance of the work to your topic and starting question
2. The timeliness or recency of the work (particularly important in scientific research projects)
3. The author of the work (based on all available information)
4. The prestige or nature of the journal (scholarly or popular press)
5. The controversial nature of the source (whether it agrees with or contradicts other sources)

As you encounter new sources, you will be the best judge of whether or not a particular source contains useful information for your research project. Record your evaluative comments in your research notebook.

✦ EXERCISE

1. Write an evaluation of one book or article you have located in your library search (or an article assigned by your teacher). Use the above criteria for evaluating the source.

Writing from Sources

Reading actively and taking accurate and careful notes in the form of paraphrases and summaries are the first important techniques for working with sources. Your reading notes will form the basis for all your subsequent writing about that particular source. In this section, we will discuss three important approaches to source books and articles that result in three different kinds of writing. These are (1) summarizing the main points of the source book or article in condensed form, (2) synthesizing the information found in two or more related sources, and (3) critiquing the information found in one or more sources.[1] These three kinds of writing differ from each other in the approach the writer takes to the source in each instance. Your purpose for writing summaries will be different from your purpose for writing syntheses or critiques. Although the source or subject may remain the same, your approach to that source or subject can change, depending on your purpose. Using different approaches to the same sources will help you to understand those sources better.

Summarizing

When summarizing, the writer takes an entirely objective approach to the subject and the source. The writer of summaries is obliged to accurately record the author's meaning. To do this, of course, the summarizer must first understand the source and identify its key ideas during active reading. Since, in general, a summary is about one third as long as the source itself, this means that two thirds of the information in the original is left out of the summary. So, what do you as summarizer eliminate? Typically, it is the extended examples, illustrations, and explanations of the original that are left out of a summary. The summarizer attempts to abstract only the gist of the piece, its key ideas and its line of argument. If a reader desires more information than that provided in a summary, he or she may look up the original.

To write a summary, first transcribe your short marginal reading notes onto a separate sheet of paper (see the outline on p. 54). Read these notes and decide what you think the author's overall point was. Write the main point in the form of a thesis statement that encapsulates the central idea of the whole article.

Thesis: `Cole thinks that there are close connections between science and art that stem from the creative spirit of humanity.`

Be sure not to use the author's words; rather, paraphrase the author's central idea in your own words. Then, by combining the thesis sentence with the marginal notes, you will have constructed the first outline of your summary. Revise the outline for coherence and logical progression of thought.

Next, write the first draft of your summary, following your outline rather than the source. Use your own words, not the words of the author, paraphrasing and condensing his or her ideas. If you want to use the author's own words for a particular passage, use quotation marks to indicate the author's exact words and insert a page reference in parentheses:

Cole observes that "a tree is fertile ground for both the poet and the botanist" (54).

In the first few sentences of your summary, introduce the source book or article and its author:

In the article "The Scientific Aesthetic" (*Discover*, Dec. 1983, pp. 16-17), the author, K. C. Cole, discusses the relationship of aesthetics and science.

Follow this context information with the thesis statement, which reflects the author's position, and then with the summary itself. Do not insert your own ideas or opinions into the summary. Your summary should reflect the content of the original as accurately and objectively as possible.

When you have completed the first draft of your summary, review the source to be certain that your draft reflects its content completely and accurately. Then reread your draft to determine whether it is clear, coherent, and concise. Next, revise the summary for style and usage, making your sentences flow smoothly and correcting your grammar and punctuation. Finally, write and proofread the final draft. Remember, your summary will recount objectively and in your own words what someone else wrote, so you should refer often to the author by name.

Synthesizing

When synthesizing, you will approach your material with an eye to finding the relationships among sources. Your purpose will be to discern those relationships and present them coherently and persuasively to your potential readers. Again, the process begins with the active reading of the sources. As you read, highlight and summarize key ideas from your sources in the margins. But instead of simply summarizing the information in one source, look for relationships between ideas in one source and those in another. The sources may be related in one or more of the following ways:

- They may provide examples of a general topic, or one source may serve to exemplify another.
- They may describe or define the topic you are researching.

- They may present information or ideas that can be compared or contrasted.

You must decide in what way or ways your sources are related. When you have decided on the relationships among the sources, write a thesis sentence that embodies that relationship. This thesis sentence should indicate the central idea of your synthesis.

Write an outline of your synthesis paper based on the organizational plan suggested by the thesis statement. This outline should articulate the relationship you have discerned among the sources. For example, if the passages you read all served to describe the same topic (perhaps life in colonial New England), the structure might look like this:

1. Opening paragraph with contextualizing information about the sources and the particular situation, life in colonial New England.
2. Thesis statement describing the relationship to be discussed: Life in colonial New England is described by historians and participants as rigid in its social structure.
3. Description 1 (based on source 1: a historical work about the New England colonies).
4. Description 2 (based on source 2: a diary or journal written by an early colonist).
5. Description 3 (based on source 3: a sermon written by a colonial preacher).
6. Conclusion: All the sources combined contribute to a description of the rigid social structure in colonial New England.

After outlining your synthesis, write the first draft of your paper. In the introductory section of your synthesis, just as in the summary, introduce the sources and their authors. Follow the introduction of sources with your thesis expressing the relationship among the sources. As you write your first draft, keep your thesis in mind, selecting from your sources only the information that develops and supports that thesis. You may want to discuss each source separately, as in the example above, or you may prefer to organize your paper to present major supporting points in the most logical sequence, using information from the sources to develop or support those points. Be sure that you acknowledge all ideas and information from your sources each time you use them in your synthesis.

Upon completion of your first draft, review the sources to be sure you have represented the authors' views fairly and cited source ideas and information properly. Reread your first draft to make sure it is organized logically and that it supports your thesis effectively. Be certain that you have included sufficient transitions between the various sections of your synthesis. Revise your synthesis for style and correctness. Finally, write and proofread the final draft of your synthesis. In general, a synthesis should give the reader a persuasive interpretation of the relationship you have discerned among your sources.

Critiquing

In the third kind of writing from sources, critiquing, the writer takes a critical or evaluative approach to a particular source. When writing critiques, you argue a point that seems important to you based on your own evaluation of the issues and ideas you have encountered in your sources. Critiques are necessarily more difficult to write than summaries or syntheses, because they require that you think critically and come to an independent judgment about a topic. However, critiques are also the most important kind of writing from sources to master, because in many research situations you are asked to formulate your own opinion and critical judgment (as opposed to simply reporting or presenting the information written by others).

As in the other forms of writing from sources, critiquing begins with active reading and careful notetaking from a source. You must first identify the author's main ideas and points before you can evaluate and critique them. Once you understand the source and the issues it addresses, you are in a position to appraise it critically. Analyze the source in one or more of the following ways:

What is said, by whom, and to whom?

How significant are the author's main points and how well are the points made?

What assumptions does the author make that underlie his or her arguments?

What issues has the author overlooked or what evidence has he or she failed to consider?

Are the author's conclusions valid?

How well is the source written (regarding clarity, organization, language)?

What stylistic or rhetorical features affect the source's content?

Other questions may occur to you as you critique the source, but these will serve to get you started in your critical appraisal. To think critically about a source, look behind the arguments themselves to the basis for those arguments. What reasons does the author give for holding a certain belief? In addition, try to discern what assumptions the author is making about the subject. Do you share those assumptions? Are they valid? It is your job to evaluate fairly but with discerning judgment, since this evaluation will be the core of your critique. Formulate a thesis that states your evaluation. Do not feel that your evaluation must necessarily be negative; it is possible to make a positive critique, a negative critique, or a critique that cites both kinds of qualities.

Write an outline of your critique, including the following:

1. An introduction of the subject you wish to address and the source article you wish to critique. Be sure to include a complete citation for the source.
2. A statement of your judgment about the issue in the form of a thesis. In that thesis statement, give your own opinion, which will be supported in the critique itself.
3. The body of the critique. First, briefly summarize the source itself. Then review the issues at hand and explain the background facts and assumptions your readers must understand to share your judgment. Use the bulk of your critique (about two thirds) to review the author's position in light of your judgment and evaluation.
4. Your conclusion, which reminds the reader of your main points and the reasons you made them.

After completing your outline, write the first draft of your critique, using your outline as a guide. Make certain that all your points are well supported with specific references to the source. Also, make certain that your main points are related to each other and to the thesis statement.

Review the source to be sure you have represented the author's ideas accurately and fairly. Reread your first draft to determine whether your thesis is clearly stated, your paper logically organized, and your thesis adequately and correctly supported. Revise your critique for content, style, and correctness. Finally, write and proofread the final draft of your critique.

Unfortunately, because critiques are subjective, it is not possible to be any more explicit in guiding your writing of them. The substance of the critique will depend entirely on the judgment you make about the source. Remember, though, that a critique needs to be well supported and your opinion well justified by evidence drawn from the source itself. In the exercises that follow, you will have the opportunity to practice writing summaries, syntheses, and critiques. It will also be valuable to write summaries, syntheses, and critiques of sources you are using in your research project as a way to better understand that topic. Do all such preliminary writing in your research notebook.

✦ EXERCISES

1. *Summary*

    Carefully read, underline, and annotate a brief article that you have encountered in your own research. Using the procedure described above, write a summary of the article. Be certain to turn in to your teacher both your summary and a photocopy of the article you are summarizing.

2. *Synthesis*

    A. Use two or more articles you have encountered in your research as the basis for an extended definition of an important concept. For example, you could write an extended definition of bird census techniques based on the explanations you find in ornithology articles.

    B. Use two or more articles to compare and contrast an idea presented by different authors. Again, using the ornithology example, perhaps two or more articles seem to disagree about whether or not a particular census technique is best. You could write a paper that contrasted their views.

    C. Use the illustrations and examples from two or more articles to describe something. For example you could write a paper that described a certain species of bird during various times of the year. For such a paper, you would cite specific cases or examples as reported by the authors, but divide your examples into categories or types of distinguishing characteristics.

3. *Critique*

    Write an evaluative critique of an article you have encountered in your own research. Remember, in a critique it is appropriate to include your opinions and experiences as well as your reactions to the article itself.

4

Primary Research Methods: Writing a Research Report

In all disciplines, the primary research methods are the customary ways in which investigators gather information and search for solutions to problems they have posed. For example, when a chemist performs an experiment in the lab, that is primary research, and when an archaeologist goes on a dig, that is also primary research. When conducting primary research, researchers are gathering and analyzing data. Secondary research, in contrast to primary research, involves studying and analyzing the primary research of others as it has been reported in books and journals. So when the chemist reads the relevant journals in the field and the archaeologist reads topographical maps of the area to be studied, they are doing secondary research. Many research projects are based on a combination of primary and secondary research methods.

PRIMARY RESEARCH IN THE SCIENCES

Lab Experiments and Reports

Central to an understanding of research in the sciences is an understanding of the scientific method. In the sciences, the method by which an experimenter solves a problem is as important as the result the experimenter achieves. Guided by the scientific method, researchers investigate the laws of the physical universe by asking and answering questions through empirical research. The scientific method begins with a scientist formulating a question and developing a hypothesis that may answer the question posed. On the basis of the hypothesis, the scientist predicts what should be observed under specified conditions and circumstances in the laboratory. Next, the scientist makes and records observations, generally using carefully designed, controlled experiments. Finally, the scientist either accepts or rejects the hypothesis, depending on whether or not the actual observations corresponded with the predicted observations. As you may have discerned from this description of the scientific method, writing plays a role at every step. A researcher must describe in great detail both the method used in the experiment and the results achieved. A report of the experimental findings is based on the laboratory notes taken during the experiment. All researchers must keep written records of their work. In the natural sciences, such records generally take the form of a laboratory notebook. The researcher uses the notebook to keep a complete, well-organized record of every experiment and each experimental variable (phenomena not constant in the experiment). The researcher must record information in a clear, easy-to-understand format so that he or she (and coworkers) will have easy access to it when it is time to draw conclusions from the experiment.

In your undergraduate science courses, you are expected to conduct scientific experiments and record your methods and results in a laboratory notebook. You may also be expected to report your experiments in a systematic way. A good lab report introduces the experiment, describes the materials and methods used in collecting the data, explains the results, and draws conclusions from those results.

Your scientific experiments in your coursework will typically be connected with a laboratory. For example, many courses in the physical sciences, such as chemistry or physics, are accompanied by laboratory sections for practical lab experimentation. However, laboratory courses still involve writing. It is important for you to realize that the scientific method employed by laboratory researchers necessitates the careful, organized, and complete presentation of methods and results through written reports.

Field Observations and Reports

In some scientific disciplines, empirical or experimental work is supplemented by field observations that occur outside the laboratory. For example, a biologist interested in moose behaviors might visit Yellowstone National Park and observe juvenile moose and their parents to determine whether maternal or paternal examples are imitated in feeding behaviors. The field experiment reported on below illustrates one of two main types of writing done by scientists—the research report.[1] (The other major type of writing is the review paper, discussed in Chapter 6.)

In research reports, scientists describe in detail research that they have conducted themselves in order to share their findings with the scientific world in general and thus advance knowledge in their field. Because it is important that scientists be able to replicate each other's research, a customary format for organizing research reports has developed that can be easily recognized by all scientists.

The standard research report format follows the scientific method that was discussed on page 7. The parts generally found in the research report are Title, Abstract or Summary, Introduction, Materials and Methods, Results, Discussion and Conclusions, and Literature or Works Cited. They correspond generally to the scientific method:

1. The scientist formulates a question and develops a hypothesis that might possibly answer the question posed. (The hypothesis is typically posed in the research report Introduction.)
2. On the basis of the hypothesis, the scientist predicts what should be observed under specified conditions and circumstances. (The prediction of expectations is typically posed in the research report Introduction.)
3. The scientist makes the necessary observations, generally using carefully designed, controlled experiments. (The detailed description of the experiment is included in the Materials and Methods section of the research report.)
4. The scientist either accepts or rejects the hypothesis depending on whether or not the actual observations corresponded with the predicted observations. (The scientist's decision to accept or reject the hypothesis is generally included in the Results section of the research report. This section is typically followed by a Discussion and Conclusion section that interprets results for the readers.)

When writing your research report, you will not necessarily write it in the order that it finally appears. In fact, you may find it easiest to write the Materials and Methods section first, since it is here that you will describe in detail the experiment itself. The Discussion and Conclusion section is probably best written near the end, after you know exactly what you want to say in your Results section. Other parts, like the Abstract and the Works Cited, can be incorporated near the end of your drafting process. Any primary data may be included in one or more appendix, included immediately after the works cited list.

The following paper by a biology student illustrates a report of a scientific field experiment, conducted for an advanced biology course in field ornithology. Notice the subheadings that clearly divide the paper into its relevant sections. The references follow the CBE format typically used in the biological sciences.

SAMPLE RESEARCH REPORT: BIOLOGY

Independent Project:
DIFFERENCES IN MORNING AND
EVENING BIRD ABUNDANCE

By:
Janene Shupe

Submitted To:
Dr. Kim Sullivan
Field Ornithology

May 27, 1993

INTRODUCTION

After I started birding, I began to hear various opinions about the best time of day to sight birds. Most people seem to be partial to observing birds in the morning, evening, or both. These differences made me wonder which opinion is correct. For my study, I decided to determine whether differences in bird abundance occur between morning and evening and if so, which time of day birds are most likely to be sighted.

To test my question, I needed to measure the abundance of birds in the morning and evening. To do this, I chose to conduct a line transect census without distance. This technique is performed by walking a preset line and recording each bird you see or hear (Sullivan, 1993). Abundance is measured by assuming that all species and individuals are equally noticeable from your study line (Johnston, 1985).

METHODS

The area I chose to study was the Canal Trail above Canyon Road in Logan, Utah. The trail borders a cement canal located on a slope with a southern aspect. Vegetation surrounding the canal and trail is comprised mostly of box elders, cottonwoods, birches and mountain shrubs, as well as several forbs and grasses. The slope above the trail is a sagebrush and bunch grass community with a few scattered junipers. Private property is located below the trail. This gives the area an assortment of habitats including pastureland, gardens, lawns, and a variety of native and introduced species. Some of the houses have feeders, adding to the diversity.

My transect line began at the trail's farthest west side at the metal gate. It ended approximately 300 yards past the easternmost bridge crossing the canal.

To begin my study, I determined the approximate time of sunrise as 6:15 A.M. and the approximate time of sunset as 8:30 P.M. (U.S. Naval Observatory, 1977). I then determined that my transect would take 45 to 60 minutes to walk. From this information, I began my morning walks at about 15 minutes after sunrise and my evening walks at about an hour before sunset. Each time of day I walked with my back to the sun. On mornings I began on the west side and in evenings I started on the east side. In this way, hoped to have similar lighting both times of day.

I conducted my studies on five days within a nine day period. I began on May 7 and ended on May 16. On each of the days I conducted my study, I recorded birds both in the morning and evening. I recorded every bird I saw or heard, including those that I could not identify.

RESULTS

After collecting my data, I totalled both the number of bird species and individual birds for each sheet within a set (a set being comprised of data from both times during the same day). With this information, I comprised a list of bird species, created line graphs, and performed statistical calculations.

In all, I observed 33 species of birds, excluding unidentified birds (Appendix A). I chose to exclude the unidentified birds from my species list since they are not very meaningful and many species may overlap. However, I did include these birds in my graphs and statistical tests.

The first graph represents differences in the number of bird species between each set (Appendix B). In every set, I observed more bird species in the morning than in the evening.

The second graph represents differences in the numbers of individual birds observed (Appendix C). The majority of these sets also have more birds in the morning. However there was an exception on the fourth set. This difference may be explained by the weather. The morning of the fourth day was windy, but the evening was still.

To test the significance of the differences in my data, I chose to use a two-sample t-test. This test is used to determine whether the difference between the averages of two independent samples are significant (Freeman et al., 1991). Since my sample size is small, I chose to use the Student Curve to determine the significance of my findings.

First, I compared the difference between the number of individuals recorded in the morning and the evening (Appendix D). I calculated a P-value of five percent, which is statistically significant. Second, I compared the differences between the number of species recorded in the morning and evening (Appendix D). This also was statistically significant, with a P-value of five percent.

CONCLUSIONS

My results showed that birds were more abundant in the morning than in the evening. This might mean that birds are more abundant in the morning and that this is the best time to observe birds. However, I do not believe that this was proved by my experiment. Many factors could have varied my results. Also, my results may be true only for specific circumstances.

First, I believe it is important to consider bias. Bias may be caused by observers, birds, habitat, weather, or censusing methods (Sullivan, 1993). I tried to avoid as much bias as possible with my methods. However, I still see a lot of problems with my project. The main sources of bias in my experiment were due to my lack of identification skills and the human activity on my site.

I probably overlooked a lot of birds that I could not identify by song. Also, I had a lot of birds recorded that I could not identify by sight. Since my skills improved each time I walked my transect, I would have guessed that my species list would have increased each time. Still, I found more species in the morning than the evening. This may mean that the differences between morning and evening might have been larger.

Another consideration is the habitat I chose. Birds may have been more active in the morning due to human disturbance. The Canal Trail is used to access other trails leading to 400 North and Utah State University. It is also used by many people for recreation. Birds in this area may be more active in the morning to avoid the increase of people during the day and evening.

This project taught me a lot about censusing and censusing techniques. I was unaware of the amount of information you can find with census studies. Census studies can be used to generate species list, estimate bird densities, discover abundance, and make numerous comparisons (Johnston, 1985). I also learned that there are many difficulties with conducting a census and that it takes a lot of time, effort, and experience to account for these.

I also learned a lot about the many techniques that can be used to conduct a census. Some techniques are very difficult, such as the variable distance line transacts that need skilled observers. Other techniques need special equipment such as mist nets and radiotransmitters. However, many techniques are easy enough to be used by nearly any observer, such as the line transect used in this study.

This project taught me a lot about birding. By observing and recording birds, I rapidly improved my identification skills and began to associate birds with their songs. I also began to notice many things about their behavior, shape, and flight patterns.

I also discovered that research studies seem to have endless results. I discovered new questions throughout my project. Related studies on bird preferences between morning and evening could focus on differences among species, seasons, weather, temperature, human activity, habitat, and numerous other factors.

REFERENCES

Freedman, David, 1991. Statistics: 2nd Edition. New York: W.W. Norton and Company.

Johnston, Richard F. (editor). 1985. Current Ornithology, Volume 2. New York: Plenum Press.

Sullivan, Kim (lecture). 1993. Field Ornithology, Fall Quarter.

United States Naval Observatory. 1977. Sunrise and Sunset Tables for Key Cities and Weather Stations of the U.S. Gale Research Company. Detroit, Michigan. No. 1296.

APPENDIX A

BIRD SPECIES LIST

1. Black-capped Chickadee
2. Mourning Dove
3. American Robin
4. Song Sparrow
5. Common Yellowthroat
6. Northern Flicker
7. Mallard
8. Empidonax (Flycatcher) spp.
9. House Finch
10. European Starling
11. Broad-tailed Hummingbird
12. Lazuli Bunting
13. Black-billed Magpie
14. Yellow Warbler
15. Black-headed Grosbeak
16. Western Tanager
17. Chipping Sparrow
18. Pine Siskin
19. Yellow-rumped Warbler
20. Brown-headed Cowbird
21. American Goldfinch
22. Ring-necked Pheasant
23. American Crow
24. Brewer's Blackbird
25. White-throated Swift
26. Northern Oriole
27. Tree Swallow
28. Cassin's Finch
29. White-crowned Sparrow
30. Wilson's Warbler
31. Rufous Hummingbird
32. MacGillavry's Warbler
33. White-breasted Nuthatch

78 *Primary Research Methods: Writing a Research Report*

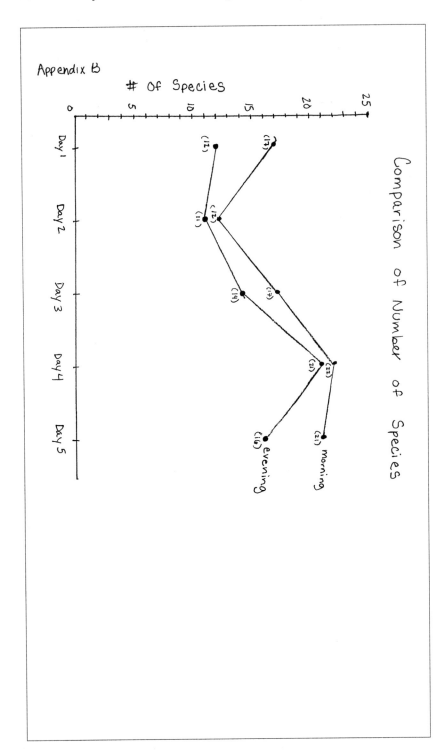

Chapter 4 / Primary Research Methods: Writing a Research Report 79

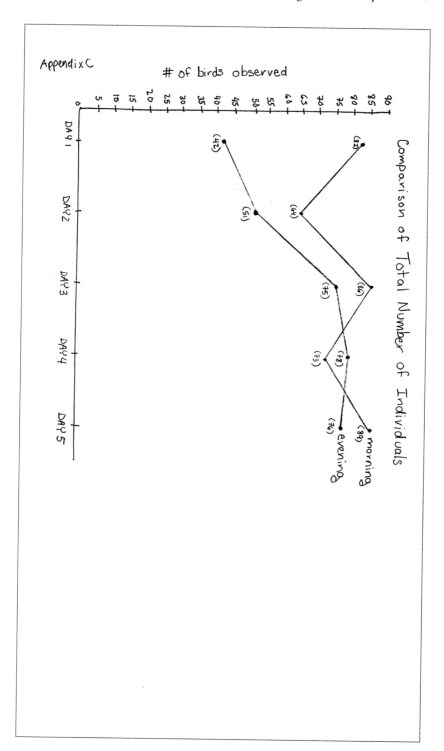

APPENDIX D

Differences Between Individuals

Morning Average 79.8 S.D. 9.2 S.E$_{sum}$ 20.6

S.E.$_{AVE}$ 4.1

Evening Average 63.4 S.D. 14.95 S.E.$_{sum}$ 33.4

S.E.$_{Ave}$ 6.7 S.E.$_{dif}$. 7.9 T = 2.1 9 = 5%

Differences Between Species

Morning Average 17.8 S.D. 3.5. S.E.$_{sum}$ 7.8

S.E.$_{Ave}$ = 1.6

Evening Average 14.8 S.D. 3.5 S.E.$_{sum}$ 7.8

S.E.$_{Ave}$ 1.6 S.E.$_{dif.}$ 2.3 T = 1.9 P < 5%

RESEARCH REPORTS IN TECHNOLOGY AND ENGINEERING

Although reports in technology and engineering have much in common with reports in the sciences, there are also some differences. In addition to the report of research, as described above, these other types of reports may be used in the sciences and technology:[2] problem analyses reports, recommendations reports, equipment evaluation reports, progress/periodic reports, and laboratory reports. The format of each report will vary depending on its purpose and audience. Most, however, follow what Pfeiffer calls the ABC format: Abstract, Body, Conclusion (212).

The following research report from an engineering course combines elements of the problem analysis and equipment evaluation reports.[3] It ends with a recommendation on a particular equipment modification that the author proposes will solve the problem.

Notice the easy-to-read outline format used by the author. This type of organizational plan helps readers to locate and read pertinent information. The works cited list follows the numbering style also commonly used in technology.

SAMPLE RESEARCH: ENGINEERING AND TECHNOLOGY FORMAT

Preliminary Design of Thermal Coupling
Switch to Use Aboard Spirit II

TO:

Professor J. C. Batty,

Cryogenics Engineer

Utah State University

FROM:

Marion D. Dart, Mechanical

Engineering Student

Utah State University

May 18, 1988

FOREWORD

When the cryogenically cooled optical instruments of the Spatial Rocket-borne Interferometric Telescope (SPIRIT II) are in operation, it is desirable to couple them directly to the cryogen coolant to prevent them from warming past 25 K. Currently, there is no coupling device in the design of SPIRIT II, and the optical instruments are allowed to warm continuously during the ten-minute flight. I was assigned by Dr. J. C. Batty to design a mechanical switch that will couple or decouple the instruments from the coolant upon demand. This paper reports the results of my preliminary design.

SUMMARY

This report presents the preliminary design of a thermal switch that will prevent the instruments aboard SPIRIT II from warming past 25 K during flight operation. When a coupling device closes the joint between the heat straps attached to the tank and the instruments, the heat entering the instrument chamber during flight is permitted to conduct directly to the cryogen tank. The size of the switch allows it to be added to the dewar system after SPIRIT II has been built. A simplified thermal analysis shows the only resistance to heat flow is the contact resistance in the joint. This resistance and mechanical efficiencies need to be determined in future testing. Based on the preliminary design, implementing the thermal switch on SPIRIT II is highly feasible.

1.0 INTRODUCTION

Optical instruments that operate in space frequently need to be cryogenically cooled. During times when the instruments are in operation, it may be desirable to couple the instruments directly to the cryogenic cooler. This is particularly advisable in the case of SPIRIT II because the instruments are cooled by the thermal conduction of heat from the instruments to the cryogen tank. The dewar system (the instruments and the cryogen tank) is made of 6061-T6 aluminum. This alloy's thermal characteristics are such that the thermal conductivity drops from 2000 w/(m*K) at 100 K to 20 w/(m*K) at 20 K (1). Because of the drop in thermal conductivity, the heat that enters the optical instrument chamber during operation has difficulty conducting to the cryogen tank before the instruments warm past 25 K.

When the instruments are allowed to warm past 25 K the risk of invalid data being collected is increased. This increase results from contaminants that collect on the walls of the dewar during cryogenic cool down being released and interfering with operation of the sensors. The telescope and interferometer were designed to function at 20 K, and a change in operating temperature of more than 5 K can result in improper performance.

During flight 20 watts of heat flood the instrument chamber and must be removed quickly if the instruments are to stay within the required temperature range. A permanent coupling device would keep the instruments at 5 K, which is 10 K too cold for proper operation. The thermal switch described in this paper is designed to couple or decouple the optical instruments of SPIRIT II directly to the cryogen tank with high thermal conductivity heat straps. The thermal switch will allow the optics to stay within the desired operating temperatures before flight and remain cool when the 20 watts of heat enter the instrument chamber by conducting it directly to the liquid helium cryogen tank. Aspects of the design covered in this report are: Constraints and Simplifications, Mechanical Components, Design Problems.

2.0 CONSTRAINTS AND SIMPLIFICATIONS IN DESIGN

SPIRIT II's structural design is complete, and no structural changes are to be made during implementation of the switch. SPIRIT II is to fly in July 1988, which allows three months for the design and implementation of the switch if it is to be used in the upcoming flight. This limitation in time calls for simplifications in the design. A simplified thermal analysis was utilized in the design of the switch.

2.1 Structural Constraints

Because no structural changes can be made to the dewar system, the thermal switch is designed so that it can be attached to the system after SPIRIT II has been built. The size of the switch is constrained so that it physically fits in the allowable space in the dewar system. Figure 1 shows the current design of the dewar system. An area of approximately 4 square inches is available on the main flange to attach the coupling device.

2.2 **Simplified Thermal Analysis**

An approach to cooling the instruments that keeps with the same conduction theory used in the current dewar system is a thermal strap made of a high thermal

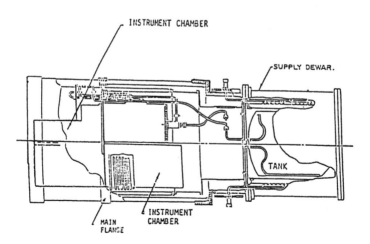

Figure 1. Current Design of Cryogenic Dewar System

conductivity metal such as copper fastened to the cryogen tank joined to another strap secured to the optical instruments. Where the thermal straps meet, a mechanical coupling device mates the straps upon demand. The thermal analysis of the switch was simplified by using lumped parameters. The thermal strap fastened to the tank was lumped with the tank by assuming it was the same temperature as the tank, 4.2 K. The strap connected to the instruments was lumped with the optical instruments and assumed to be 20 K.

The assumptions of lumped parameters employs the fact that the straps are to be made of 99.99% pure copper. Copper with this purity has a thermal conductivity of 3000 w/(m*K) between 4 and 20 K (2). A thermal conductivity of this magnitude allows the resistances to thermal conduction in the heat straps to be neglected. With the impedance in the straps neglected, the only resistance in the conduction path is the contact resistance between the two straps when the coupling device closes the joint.

3.0 MECHANICAL DESIGN

The design of the switch consists of two parts: the design of the coupling device and the design and fastening technique of the heat straps.

3.1 Coupling Device

The coupling device is to ensure a junction between the straps connected to the tank and the ones connected to the instrument chamber. This joint must be closed with enough force to ensure minimized contact resistance between the straps. Because of the small area (4 square inches) allowed for attaching the coupling device, only a small motor can be used. A significant mechanical advantage is needed if the device is to perform as required. A commercially available motor was found that can supply 5 oz.-in. torque. This available torque is used to supply 50 lbs. force with the mechanism pictured in Figure 2. Appendices 1a. through 1c. describe the mechanism in more detail.

The motor pinion in Figure 2 drives the gear attached to the ball screw with

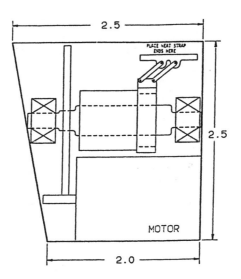

Figure 2. Coupling Device Mechanism Used in Switch

an 18:1 gear ratio. When the ball screw turns, the ball nut moves linearly causing the linkage connected to the nut to compress a spring. When the spring is compressed, 50 lbs. is applied to the ends of the heat straps in place on the device forcing them into thermal contact. An indium gasket is between the adjoining straps to help ensure good thermal contact.

3.2 Heat Straps

The heat that floods the instrument chamber during operation will be conducted to the cryogen tank by means of heat straps connected to the tank and the instruments. These straps are joined upon demand by the coupling device. The heat straps needed to be made of a high thermal conductivity metal so the heat can flow quickly to the cryogen tank without allowing the optical instruments to warm more than 5 K. A high-purity copper was chosen as the best metal because of its availability, malleability, and thermal conductivity.

3.2.1 <u>Heat Strap Design</u>. The heat straps are designed to have as large a cross-sectional area as will fit in the available space. This will increase the validity of the lumped parameter heat transfer analysis by providing a large thermal mass. The straps fastened to the cryogen tank will have a cross section of 1.25 in. × 0.375 in. There will be two straps connected to the tank, each formed to run along either side of the telescope into place on the coupling device mounted on the main flange. The heat straps secured to the optical instruments will be flexible copper braids made of two hundred 20 AWG individual copper wires braided together. Two braids will be wrapped circumferencially around the instrument chamber with an end of each braid run into place on the coupling device.

3.2.2 **Fastening Technique**. The heat straps connected to the tank are to be bolted securely to the tank as described in Appendix 2. Between the copper straps and the helium tank, an indium gasket must be used. This gasket will reduce the contact resistance between the strap and the tank making it negligible. The heat straps that are attached to the instrument chamber are made from a flexible copper braid. The braid will be wrapped around the chamber, and metal banding will secure the braid in place. It is important to attach the banding as tight as possible and use an indium gasket between the chamber and the copper braid to minimize the contact resistance.

4.0 UNRESOLVED DESIGN ASPECTS

The short time allowed for this preliminary design has left parts of the design unfinished. Before the switch is built, thermal properties of materials used in the switch and the efficiency of the coupling mechanism at cryogenic temperatures must be determined. An extensive cost analysis will also need to be done.

4.1 Required Testing

To ensure reliability of the thermal switch to maintain the optical instruments within the desired temperature range of 15 K to 25 K, tests that measure the contact resistance in the joint and evaluate the mechanical efficiencies of the coupling device should be run.

4.1.1 Contact Resistance in Joint.
The thermal contact resistance in the joint where the adjoining ends

of the heat straps meet can only be determined by testing. Because this is the only resistance the heat flowing from the instrument to the tank sees, it is vital that the force in the coupling joint be great enough to overcome that resistance. Based on tests described in M. Baugh's lab book (3), 50 lbs. of force will overcome the contact resistance in the joint and allow 20 watts to flow to the tank. Proper tests can verify this assumption or determine the necessary force to overcome the contact resistance. Modifying the spring can produce more or less force.

4.1.2 **Mechanical Efficiencies**. The efficiencies of the motor, ball screw and ball nut at cryogenic temperatures must also be determined. This can be done by testing the assembly at design conditions.

4.2 **Cost Analysis**

To date no concern has been given to the cost of the switch. Many of the parts, however, are commercially available, which should reduce the initial cost. An extensive analysis should include testing costs and labor costs along with the prices of individual components of the switch.

5.0 CONCLUSIONS

This report presents the preliminary design of a mechanical thermal switch. When the cryogenically cooled optical instruments of the SPIRIT II rocket are in flight operation 20 watts of heat enter the instrument chamber. If this heat is not removed within seconds, the instruments will warm past 25 K. At temperatures above 25 K the instruments are more likely to function improperly. A thermal switch that couples the instruments to the cryogen coolant will prevent the instruments from warming past the desired temperature. A preliminary design of the switch consists of the following:

1. Size constraints that allow the switch to be attached to the dewar system after SPIRIT II has been built

2. A cooling process that uses conduction heat transfer

3. Simplified thermal analysis using lumped parameter theory

4. Heat straps that are fastened to the instrument chamber and heat straps that are connected to the cryogen tank

5. A coupling device that joins the heat straps connected to the instrument chamber and tank to provide thermal conduction

6. Description of required tests to ensure switch reliability

    a) Contact resistance test

    b) Mechanical efficiency test

Based on the preliminary design, I feel that implementing a coupling switch on SPIRIT II is quite feasible.

Works Cited

(1) National Bureau of Standards. <u>Handbook of Materials for Superconducting Machinery: Mechanical, Thermal, Electrical, Magnetic Properties of Structural Materials</u>. New York: Metals and Ceramics Information Center, Advanced Research Projects Agency and Cryogenics Division, 1977.

(2) Touloukian, J. <u>Thermal Physical Properties of Matter</u>. New York: IFI/Plenum, 1970.

(3) Baugh, M. <u>Senior Design Log Book</u>. Department of Mechanical Engineering, Utah State University, Logan, Utah, 1987.

98 *Primary Research Methods: Writing a Research Report*

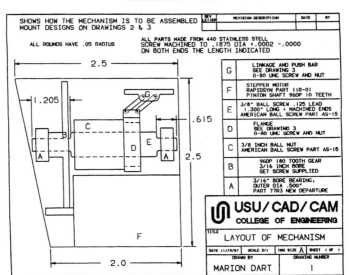

[Student also included additional appendices: 1b-Mechanism Mounts; 1c-Bracket and Linkage; 2-Cold Straps on Tank]

NOTES

1. I am grateful to Dr. Kim Sullivan of Utah State University's Biology Department for providing me with this student paper.

2. This outline of technical report formats is taken from William S. Pfeiffer. *Technical Writing: A Practical Approach,* 2nd ed. New York: Merrill, 1994.

3. I am grateful to Professor Batty of Utah State University's College of Engineering and Nancy O'Rourke of Utah State University's English department for providing me with this student paper.

5

Planning, Writing, and Revising Your Research Paper

After you have completed the primary research, library research, and preliminary writing on your topic, you are ready to begin planning the actual research paper. You should consider carefully the following two important components as you begin to plan: rhetorical situation and organization.

RHETORICAL SITUATION

The context in which you are writing an assignment is called the rhetorical situation. The term rhetoric refers to written or spoken communication that seeks to inform someone of something or to convince someone of a particular opinion or point of view. For any writing assignment, you need to analyze the components of the rhetorical situation: (1) the writer's purpose, (2) the writer's persona, (3) the potential readers or audience, (4) the subject matter, and (5) the appropriate language or tone.

Purpose

When preparing to write, a writer must decide on the actual purpose of the piece. What is the goal that should be accomplished? Many times the goal or purpose is implicit in the writing task itself. For example, for a newspaper reporter, the goal is to present the facts in an objective manner, describing events for newspaper readers. For your research paper assignment, you need to determine your purpose or goal and define it carefully. The purpose does not have to be grandiose or profound—it may simply be to convince your readers that you have a fine grasp on the topic and are making some important points, or it may be to inform your readers of the current state of knowledge in a particular field.

Persona

You also need to decide just how to present yourself as a writer to those who will read your work. Do you want to sound objective and fair, heated and passionate, sincere and persuasive, or informative and rational? The term persona is used to describe the identity that the speaker or writer adopts. As you know, we all play many roles, depending on the situations in which we find ourselves: with our parents, we may be quiet and reserved; with our peers, outgoing and comical, and so on. Similarly, you can be flexible about how you portray yourself in your writing, changing your persona with your purpose and audience. First, establish your credibility by being careful and thorough in your research and by showing that you have done your homework and understand what you are writing about. Then, prepare your finished product with care and attention to detail. If you do not, your readers will assume that you are sloppy and careless and will largely discount anything you have to say. Many job applicants never even make it to the interview stage because their letters of application convey the subliminal message "here is a person who is careless and inconsiderate of others."

Audience

Identifying those who may be reading your writing will help you to make decisions about what to include or not include in your research paper. Those who are the most likely to read your writing make up your audience. For example, a newspaper reporter assumes a general readership made up of members of the community. But the reporter must also assume that his or her readers were not present at the event being covered; thus, he or she must take care to reconstruct details for the readers.

In the case of a college research paper, the instructor of the course may suggest an appropriate audience or may in fact be the primary audience. You will wish to discuss the paper's potential audience with your teacher. If the instructor is the intended audience, you should assume that the instructor is knowledgeable about the subject, reasonably intelligent, and particularly interested in the accuracy of the research.

Take time before writing to consider carefully who will read your research paper. Your readers make a difference to you, both in how you approach your topic (Are my readers novices or experts in the field?) and in the tone you adopt (Are my readers likely to agree with me or must I win them over to my point of view?). If your instructor has made no stipulations about the intended audience for your research paper, you should discuss the issue of audience with him or her. If your teacher is the target audience, it is especially important for you to know whether or not your teacher will be reading as a nonexpert (a novice in the field) or will assume the role of a knowledgeable expert. Your decisions about what to include in the paper and what level of tone and diction to adopt will depend on whether you are writing for an expert or novice audience.

Subject Matter

The most important component of the rhetorical situation, however, is the subject matter. Although no piece of writing exists in isolation (hence the need for analyzing the purpose, persona, and audience), the content that you are presenting to your audience will be the core of any piece you are writing. You must decide from the mass of material you discover in your research what to include in your written presentation. These decisions are based on your starting question, your analysis and evaluation of your sources, and your thesis statement. Knowing what your ultimate goal is, how you wish to sound, and who your readers are will help you decide what source materials to use.

Appropriate Language or Tone

Knowing your purpose, persona, audience, and subject matter will lead you also into appropriate decisions about language and tone. If your purpose is to inform a general audience on a technical subject, you will need to take particular care with defining terms and using general words in place of technical jargon. You might also consider providing your readers with a glossary of terms to help them through technical information.

If you are addressing an audience of specialists, your rhetorical decisions about language and tone will be quite different. In this case you may wish to use technical vocabulary to identify yourself with the expert audience and to build your own credibility. You will need to take care not to bore an expert reader by providing too much background information, which such readers will not need.

Organization

Once you have gathered and evaluated source materials on your topic, completed your preliminary writing assignments, and analyzed the rhetorical situation, you can begin to organize your ideas. During the planning stages, you need to decide how you will give pattern and order to your research paper. The importance of planning cannot be overemphasized. Readers will use your skeletal plan, which should appear in some form in the written research paper, to reconstruct your meaning. Many recent investigations into the reading process have shown that readers reconstruct meaning in writing by using organizational plans, that is, explicit directional signals left by the writer in his or her work.

There is no one right way to make order out of the mass of material you have gathered in your research. Some people find that just beginning to write helps them to "discover" a direction and a pattern. Others prefer to outline and to organize or sort their notes by categories. Some writers spread their notes in seemingly random piles across the desktop; others sort notecards neatly by topics. Eventually, regardless of the process, you want to be able to write a thesis statement that captures succinctly the main point you wish to make in your paper.

REVISING YOUR RESEARCH PAPER

Important work still remains on your paper once you have completed a rough draft. You must revise the paper to make the most effective possible presentation of the research. Your readers expect you to be clear and correct in your presentation so that they are not distracted by confusing language or incorrect punctuation. Read your rough draft several times, both on the computer screen and on hard copy. Each time you read it, pay attention to a different aspect of the paper for possible revision and correction. At first, when reading your rough draft pay attention to the overall structure and style of the paper; the second time through, check grammar and punctuation; the third time, make sure source materials (paraphrases and quotes from sources) are incorporated smoothly and accurately into the text. Finally, consider formal details such as conventions of documentation, format, and presentation. After your paper has been typed and spell-checked, proofread it several times to catch and correct all typographical or transcription errors.

Revising for Structure and Style

The first time you read your draft, pay attention to the organizational structure and overall content of the paper. At this point, decide whether you need to make any major changes in the order of the ideas or whether you should alter the tone. Use the power of your word processing program, in particular its cut and paste functions, to accomplish quickly and easily these global changes. A helpful acronym to keep in mind as you revise for such large issues is EARS: *E*liminate, *A*dd, *R*earrange, and *S*ubstitute.

Do not be afraid to eliminate irrelevant material from your paper. Your teacher will prefer a paper that is tightly focused to one padded with irrelevant details. Conversely, if you discover a section of your paper that seems thin, do not hesitate to add more information: more evidence to support an idea, more explanation to clarify an idea, and so on. Be sure that the major sections of the paper are arranged in a logical order. If there seems to be any confusion, rearrange major sections. Finally, if you find an example or a piece of evidence that does not seem persuasive in the context of the paper, substitute a new example for the one you currently have.

Remember, your rough draft is exactly that—rough. It is important for you to read it critically now so that you can improve the overall presentation of your ideas. Share your draft with a friend, spouse, teacher, classmate, or coworker. Another reader can provide insights to your paper that may be very helpful to you as your revise. The following list of questions can guide both you and others as they read your draft:

1. Is my title accurate and does it relate to the paper?
2. Is my introduction complete? Does it provide necessary background information?
3. Is my thesis clearly stated early in the paper so that the reader knows what to expect?
4. Is the organizational pattern of the paper clearly marked for the reader by subheadings or directional signals?
5. Are the various sections of the paper linked by good transitional words and phrases?
6. Have I used a variety of evidence to convey my points: examples, analyses, primary or secondary data, analogies, illustrations, narratives, descriptions?
7. Do the sections of the paper appear in a logical order or do I need to rearrange the parts? Does the logic of the argument seem clear?
8. Is each section of the paper supported with sufficient data and evidence from the sources and from my primary research? Are

sources integrated smoothly into the flow of the paper? Can the reader tell where sources stop and my own writing begins?
9. Is my conclusion adequate? Does it return to the thesis and highlight the answer to the starting question that motivated the research?
10. Have I demonstrated a depth of analysis and complexity of thought concerning the topic such that readers will feel significantly taught by me?

Rework any troublesome aspects of the paper. If a particular section of the paper lacks sufficient evidence, go back to the library for some appropriate supporting material. If you are unsure about the tone of your paper, the clarity of the language, or the presentation of your ideas, ask for specific advice from other readers. Such outside readings of your work can sometimes provide the distance needed for an objective evaluation.

You may wish to try the technique of reverse outlining as a way of seeing any global structural problems in your draft. To reverse outline, number each paragraph in your draft. Then summarize in one sentence the essential content of each paragraph. In this way you may discover sections out of sequence or paragraphs on the same topic many pages apart. Or, if you are having difficulty summarizing a particular paragraph, you may find you have tried to cover too much information and need to divide a longer paragraph into a series of shorter, more focused paragraphs.

Improving Paragraphs

As previously suggested, you first look at the overall structure and content of your paper. Then begin to narrow the scope of your revising to individual paragraphs. Check to be sure that each paragraph has a single major focus and that the ideas within the paragraph are all related to that focus. Focusing your paragraphs in this way is a great aid to your reader. When this is done, a new paragraph indicates a change to a new idea or change in direction. Often, the first sentence of each paragraph serves as a transitional sentence, bridging the gap between the ideas in the two separate paragraphs. This is the time to check for transitions between paragraphs as well as paragraph focus. As you revise your paragraphs, ask yourself the following questions:

1. Does each paragraph relate to the overall point of my paper?
2. Does each new paragraph contain its own internal focus or coherence?

3. Does the first sentence of each paragraph offer a bridge or transition from the previous paragraph?
4. Is the language used in each paragraph concrete and clear? Are there unnecessary words or phrases that I should delete?
5. Is the tone of each paragraph objective? Do I sound interested and concerned about the subject but not overly emotional?

Improving Sentences

In continuing to narrow the scope of your revising, look at individual sentences within your paper. Revise any sentences that seem awkward or confusing. In general, the more simply and directly you state your ideas, the better. Do not use overly complex sentence structures—they will only confuse your reader. Any very long sentences may need to be broken down into shorter sentences. On the other hand, a series of short, choppy sentences may be more effective if rewritten as a single long sentence. Reading your draft aloud—to yourself or to someone else—can often help you to "hear" problematic sentences. Or perhaps your teacher can make additional suggestions about revising your sentence style.

Improving Words

Careful attention to wording in your paper will help you to most accurately report your research findings or review the literature for your readers.

Look at individual words in your paper with an eye toward spotting confusing vocabulary or unnecessary jargon. Define any terms that might be unfamiliar to a general reader and replace any jargon specific to a field or discipline with more common words. As a general rule of thumb in science writing, be simple and be concise. If you can write your findings in simple and direct language, it is a sure sign that you have come to grips with your subject and understand it thoroughly yourself. Often language that seems obscure reflects an underlying muddiness in the writer's thinking.

Verbs

A conventional usage for verb tenses has grown up over the years in science writing. When reporting your own findings, use the *past* tense; when reviewing the published work of others, use the *present* tense:

I *observed* the birds for twenty-four hours. Sturkie (1986) *shows* that fat content of the birds *varies* by age, sex, species and nutrition.

In much science writing the intended clarity of expression may be lost behind empty verbs and passive constructions. Verbs are one part of speech that often cause science writers problems. Performing these few simple editing tasks with the verbs in your paper will vastly improve your own writing.

Content Verbs versus Empty Verbs

The verb "to be" (is, was, were, am, are) asserts a state of being, telling us only that something exists. Because the "to be" verb in its various forms essentially has become empty of meaning, you should attempt to replace "to be" verbs with content verbs that do convey meaning.

EXAMPLES:

Original—with empty "be" verb

> It is the custom for visitors to remove their shoes before entering a Japanese home.

Revision—with content verb

> Visitors customarily remove their shoes before entering a Japanese home.

Original—with empty "be" verb

> Many planes are twenty years old and will have passed their life expectancy of ten to fourteen years.

Revision—with content verb

> Now twenty years old, many planes have long since passed their life expectancy of ten to fourteen years.

Action Verbs versus Nominalizations

Many writers slow down their readers by using complex nouns in their sentences instead of the more active verbs that those nouns come from. For example, *decision* is the noun form (nominalization) of the verb *to decide*, and *invasion* is the noun form of *to invade*. Somehow writers have gotten the erroneous impression that the use of nominalizations makes their writing sound more important or official. Changing your verbs into nouns, however, robs them of their power and motion, thus slowing down the reader's progress. Whenever possible, change nominalizations to action verbs.

EXAMPLES:

Original—Sentence with nominalization
   This land has the appearance of being arid.

Revision—Sentence with action verb
   This land appears arid.

Original—Sentence with nominalization
   He finally came to his decision.
   He would run for office.

Revision—Sentence with action verb
   He finally decided to run for office.

Active Voice versus Passive Voice

The passive voice can be used effectively in writing when we either do not know who the subject is or don't want the subject known, as in the following:

Passive: The Vietnamese countryside was bombed. [by whom?]
Active: The U.S. Air Force bombed the Vietnamese countryside.
Passive: The boys were asked to leave. [by whom?]
Active: The neighbors asked the boys to leave.

Although passive voice has a legitimate function, it is often overworked in writing. To keep the pace of your writing moving along and to provide your readers with often essential information, try to use the active voice whenever possible.

Commonly Misused Words

The following words are frequently misused in science writing.

Due to: Correctly used following a subject to attribute something to that subject.

   Correct usage: The kestrel's decreased weight is *due to* its lack of food during winter.

   Incorrect usage: *Due to* its lack of food, the kestrel decreased in weight.

Compare:

1. Followed by *to* when comparing two similar people or things or implying an analogy.

Correct usage: He *compared* the small brown bats *to* a hibernating bear.

2. Followed by *with* when comparing details of very similar or very dissimilar people or things.

Correct usage: When compared *with* eagles, kestrels are more efficient fat-burners.

Data: Data is the plural form; *datum* is the singular.

Different from: preferred usage; (*different than* is the less-preferred option)

Index: Index is the singular form; the plural form is *indexes* as in book indexes, or *indices* as in measurable quantities.

-like: When used as a suffix, a compound word is generally formed (lifelike, birdlike). When the suffix follows a word ending in a double 'l' or a long compound word, use a hyphen (shell-like, mammalian-like).

mucous: adjective (mucous membrane).

mucus: noun (mucus = a viscous solution)

over/under: not to replace *more* or *less than* when describing quantities.

Correct usage: We found *more than* 60% deviation from the norm.

Incorrect usage: We found *over* 60% deviation from the norm.

Parameter: Has a special meaning in mathematics and statistics; do not use loosely for variable, quantity, quality, determinant, or feature.

Percent: May be a noun, adjective, or adverb. When used with numbers, use the symbol. Correct: 98%. Incorrect: 98 percent.

Percentage: Noun, meaning part of the whole as expressed in hundredths, as in the *percentage* of animal cells.

Correct usage: *percent* error (adjective)

Incorrect usage: *percentage* error

Significant/significance: In science papers, confine the use of this term to statistical judgment; do not use loosely for important, notable, or distinctive.

Editing for Grammar, Punctuation, and Spelling

Once you have revised the overall structure and style of your paper, you are ready to read the paper again, this time with an eye to grammar, punctuation, and spelling. It is important to present your ideas clearly, but it is equally important to present your ideas correctly. A reader will discount you as either ignorant or careless if your work is full of grammatical errors.

As you read your paper again, ask yourself the following questions:

1. Is the grammar of each sentence correct? Does each sentence contain subjects and predicates?
2. Do subjects and verbs agree?
3. Are pronouns clear and unambiguous in their reference?
4. Is the overall punctuation correct?
5. Have I punctuated and cited quotes and paraphrases correctly?
6. Are there any words that I need to look up in the dictionary, either for meaning or for spelling?
7. Have I used the active rather than the passive voice? Are my verbs vivid action words rather than empty "to be" verbs or nominals?
8. Have I varied sentence style and avoided repetition?
9. Does each sentence flow smoothly without being awkward, wordy, or confusing?
10. Is the format of my paper correct, including the title page, body of the text, endnotes, or bibliography page?

Mechanical Errors

It would be helpful for you to refer to a recent grammar and usage handbook for questions of English grammar, punctuation, and syntax. Use a dictionary to check the meaning of individual words and run spell-check with each draft. Take the time now to look carefully for all problems in your writing. As always, errors in grammar, punctuation, syntax, and spelling detract from your message and make a negative impression on your reader. A paper full of grammatical or spelling errors signals to the reader that the authority of the writer, and hence the authority of the research, is questionable. Listed below are a few of the most commonly made mechanical errors in science writing:

1. Punctuation

Commas

   a. Use a comma to separate long introductory material in a sentence from the rest of the sentence:

   Because of increased serum cholesterol levels, arteries within the circulatory system may form fatty lesions.

   b. Use commas to set off interrupting material from the rest of the sentence:

This deposit of plaque, and the cellular proliferation of smooth muscle, can become a large bulge in the artery.

c. Use commas to separate items in a series:

Serum cholesterol measures may predict the risk for infarction, death from coronary heart disease, and all-cause mortality.

d. Use a comma to separate independent clauses joined by a conjunction (and, but, or, for, nor, yet):

This risk is more apparent in cardiac patients, but it is also apparent in healthy subjects.

DO NOT use a comma to separate two independent clauses WITHOUT a conjunction:

Incorrect: This risk is more apparent in cardiac patients, it is also apparent in healthy subjects.

Semicolons

Semicolons are used in the following three cases:

a. To join two independent clauses:

This risk is more apparent in cardiac patients; it is also apparent in healthy subjects.

b. To connect two independent clauses joined with a conjunctive adverb (however, nevertheless, moreover, furthermore):

The benefits of high density lipoproteins is well documented; however, some have questioned the effects of very low levels of cholesterol.

c. As a "super-comma" in a series already containing commas:

The patients were given low-fat diets of complex carbohydrates; lean meats, fish, and poultry; and fat-free dairy products.

Colons

The main use for the colon is to introduce items in a series:

Their diet consisted of the following items: lean meats, fish, poultry, and grains.

The colon may also be used to introduce a direct quote:

In the Framingham study, the following is stated: "A relationship between cholesterol levels and prognosis in men and women does exist" (Jones 1992, p. 55).

2. Capitalization

a. Use capitals for the scientific names of a Phylum, Class, or Family:

Most fish species are included in the group Teleostei.

b. Use capitals for the complete common name of a species of bird:

Only three species, the Ash-throated Flycatcher, Scrub Jay, and Plain Titmouse, consistently provide high counts in the sampling period.

3. Italics (or underline)

a. Use italics for the scientific name of a genus, species, subspecies or variety:

Herring (*Clupea pallasii*) sperm are inactive in sea water.

b. Use italics to indicate foreign words and phrases:

ad nauseum, in vitro, et alium

c. Use italics to indicate a word or phrase you wish to emphasize:

The depth at which a subject is placed into the chamber *can* alter your results.

4. Abbreviations

Typically, in science writing terms of measurement in the text are abbreviated if they are preceded by a number. The same symbol is used for either singular or plural and no period is used:

1 hr (one hour)

12 hr (twelve hours)

3 l (3 litres); m = meter, min = minute, s = second, wk = week, year = yr, month = mo, day = da or d.

Do not abbreviate the name of a unit of measure that follows a spelled-out number, as at the start of a sentence:

Ten liters were required for proper dosing.

A genus name can be shortened to the first letter of the genus if the name has been presented earlier in the paper:

We surveyed adult Nile crocodiles (*Crocodylus niloticus*) and two crocodiles from southern Mexico (*C. acutus* and *C. moreletii*).

5. Parentheses

a. Use parentheses to indicate the scientific name when the common name of a species is mentioned for the first time.

b. Use parentheses to refer to information in a table of figure in the text: (see Figure 1) or (Table 2).

c. Use parentheses to enclose a comment or explanation that is structurally independent of the rest of the sentence:

Roby (1989) also used total body electrical conductivity (TOBEC) on the birds in his study.

    d. Use parentheses to label enumerations within a paragraph:

(1) (2)

    e. Use parentheses for internal citations:

Fat content varies with age, sex, species and nutrition (Sturkie, 1986).

    f. Rather than using two parentheses, combine them as follows:

Researchers studied the White Tern (*Gygis alba*; Dorward, 1963).

6. Brackets

Use brackets to enclose information within parentheses:

Migratory birds use fat as their main energy source (twice as much as carbohydrates [Schmidt-Nielsen, 1990]).

7. Numbers

    a. A decimal number of less than one should be written with a zero (0) preceding the decimal point:

Twenty-five birds had a narrow range of weight and their r^2 value was much lower, 0.676.

    b. Spell out a number that occurs at the beginning of a sentence:

Fifteen of the birds were weighed with the feathers removed.

    c. Numbers that modify an adjective should be written as a whole number and separated from the adjective with a hyphen:

The weight of the male kestrels was monitored for a 5-year period.

    d. Numerical values of sample size are customarily identified by the letter n (or N):

Each observation period (N = 43) was treated as a single observation, based on the two hours over which Rosa collected data.

    e. Day of the month typically precedes the month. No commas are used:

We monitored the male birds at site A from 10 July 1994 to 10 January 1995.

In most situations use words for numbers one through nine and numerals for larger numbers. Treat ordinal numbers as you would cardinal numbers: third, fourth, 33rd, 54th.

REWRITING YOUR PAPER USING WORD PROCESSING

Using a computer word processing program to revise your research paper has many advantages. Using word processing, you can quickly and easily create a paper that is as attractive and free of distracting errors as possible. Some writers compose their research papers directly at the computer, perhaps using the research database they have already created in their computer research notebooks. Others type a rough draft into the computer for revising. Using word processing makes changing your paper easier, since it allows you to eliminate, add, rearrange, and substitute material, altering individual words, sentences, paragraphs, and even whole sections of the paper without having to recopy or retype.

Revising with Word Processing

Word processing allows you to move material in your paper from one location to another. Before doing so, however, be sure you have made a backup copy of your file so that you don't inadvertently lose part of your text. Once you have moved the material (typically using the cut and paste functions), be sure to read the revised version very carefully and make any changes needed to smoothly integrate the new material into the existing text. When revising for structure and style or to improve paragraphs, sentences, and words, you will find your word processing program's text manipulation features your greatest assets.

Editing with Word Processing

Word processing also makes editing for grammar, punctuation, and spelling easier. Most word processing programs now offer their users a spell-check feature, which is extremely helpful in identifying typographical errors. You should get in the habit of running spell-check several times as you are drafting your paper. Remember, though, that the spell-check feature does not identify words that you have inadvertently misused, such as homonyms: *there* for *their*, or *its* for *it's*, and so on. You still need to proofread very carefully yourself, even after using spell-check.

Another helpful tool for editing is your word processor's thesaurus. If you find you are overusing a particular word, you can use the thesaurus to supply alternatives. Or perhaps you find the verbs you have used are not as vivid and active as they could be; the thesaurus can suggest other verbs with similar meanings. Again, a note of caution: don't use a word suggested by a thesaurus that you don't know, because it might have connotations or slight variations in meaning that do not make sense in the particular context of your writing.

Grammar-Checking Software

Special software programs are now available that can help you edit and proofread your paper for selected stylistic or grammatical problems. For example, *Grammatik* and other similar programs will identify excessive use of inactive "to be" verbs, overuse of prepositions, vague words or jargon, and so on. However, the so-called grammar-checking programs currently available can be misleading to writers. The kinds of writing problems that such programs can check are very limited. For example, they are virtually useless at checking punctuation because they are not sophisticated enough to really analyze the underlying grammar of each sentence. Only you, the writer, can do that. So, if you use a grammar-check feature, be aware of what it can and cannot do.

Bibliography and Footnote Software

Many word processing programs currently on the market offer features to help you generate the footnotes, endnotes, or bibliography for your paper. I have found these features only marginally useful in my own writing, but I know some writers use them quite extensively. If you are using footnotes, having the word processing program automatically place them at the bottom of the appropriate page can be a big help. However, the documentation software is only as good as the information you give it, that is, you still need to type in to the computer all of the bibliographical information. The software then manipulates the information, placing it in the proper order and formatting it appropriately. You need to be aware that the documentation style built into the software may not be the same style as you need to use for your discipline. Again, you should be the final judge of what is the correct documentation format for your paper.

If you have access to computer word processing, you may want to investigate some of the software previously discussed that is designed to help you with your writing. But do not take a computer program's advice as gospel; you cannot count on a computer to know what is best for your own writing; only you can know that.

Incorporating Reference Materials

The earlier section on writing the rough draft suggested that you not worry about the smooth incorporation of quoted material until you began revising your paper. We have now reached the stage at which you should double-check all of the source material in your paper to be sure you have incorporated it smoothly and appropriately into the flow of your ideas. Your source material should be primarily in the form of paraphrases and summaries that you use to reinforce your own points. Even though they are written in your own words, both paraphrases and summaries still require documentation (identification of the source either through in-text citations or footnotes or endnotes). Putting source material into your own words greatly improves the flow of your paper, because the paraphrase style will blend with your own writing style and thus be consistent throughout.

You should use direct quotation very sparingly. Reading strings of direct quotations is extremely distracting; using excessive quotation creates a choppy, disjointed style. Furthermore, it leaves the impression that you as a writer know nothing and are relying totally on what others have said on your topic. The better alternative is to incorporate paraphrases and summaries of source material into your own ideas both grammatically and logically. At this time, check your paper to be sure you have documented all source material accurately and fairly. By following the documentation style in Chapter 6 you will be able to produce a paper that is correctly and accurately documented for your chosen academic discipline. In general, remember to document both completely and consistently, staying with one particular documentation style.

Incorporating Direct Quotes

At times you may want to use direct quotes in addition to paraphrases and summaries. To incorporate direct quotes smoothly, observe the following general principles. However, it would be wise to also consult the style manual of your discipline for any minor variations in quotation style.

1. When your quotations are four lines in length or less, surround them with quotation marks and incorporate them into your text. When your quotations are longer than four lines, set them off from the rest of the text by indenting five spaces from the left and right margins and triple-spacing above and below them. You do not need to use quotation marks with such block quotes. (Note: In some disciplines, block quotes are customarily indented ten spaces from the left margin only and double-spaced throughout.) Follow the block quote with the punctuation found in the source. Then skip two spaces before the parenthetical citation. Do not include a period after the parentheses.
2. Introduce quotes using a verb tense that is consistent with the tense of the quote. (A woman of twenty admitted, "I really could not see how thin I was.")
3. Change a capital letter to a lower-case letter (or vice versa) within the quote if necessary. (She pours her time and attention into her children, whining at them to "eat more, drink more, sleep more.")
4. Use brackets for explanations or interpretations not in the original quote. ("Evidence reveals that boys are higher on conduct disorder [behavior directed toward the environment] than girls.")
5. Use ellipses (three spaced dots) to indicate that material has been omitted from the quote. It is not necessary to use ellipses for material omitted before the quote begins. ("Fifteen to twenty percent of anorexia victims die of direct starvation or related illnesses . . . [which] their weak, immuneless bodies cannot combat.")
6. Punctuate a direct quote as the original was punctuated. However, change the end punctuation to fit the context. (For example, a quotation that ends with a period may require a comma instead of the period when it is integrated into your own sentence.)
7. A period, or a comma if the sentence continues after the quote, goes inside the quotation marks. (Although Cathy tries to disguise "her innate evil nature, it reveals itself at the slightest loss of control, as when she has a little alcohol.") When the quote is followed with a parenthetical citation, omit the punctuation before the quotation mark and follow the parentheses with a period or comma: Cal has "recognized the evil in himself, [and] is ready to act for good" (Cooperman 88).
8. If an ellipsis occurs at the end of the quoted material, add a period before the dots. (Cathy is "more than Woman, who not only succumbs to the Serpent, but becomes the serpent itself . . . as she triumphs over her victims. . . .")
9. Place question marks and exclamation points outside the quotation marks if the entire sentence is a question or an exclamation. (Has Sara read the article "Alienation in *East of Eden*"?)

10. Place question marks and exclamation points inside the quotation marks if only the quote itself is a question or exclamation. (Mary attended the lecture entitled "Is Cathy Really Eve?")
11. Use a colon to introduce a quote if the introductory material prior to the quote is long or if the quote itself is more than a sentence or two long.

    ```
    Steinbeck puts it this way:
    [long quote indented from margin]
    ```

12. Use a comma to introduce a short quote. (Steinbeck explains, "If Cathy were simply a monster, that would not bring her in the story.")

Formatting and Printing Using Computers

Besides helping you as a writer, word processing can also help you create a text that is professional in appearance. However, attention to format should be the last consideration of your writing process. Too often, writers using word processing spend excessive amounts of their time playing with the appearance of the text, varying the fonts, for example, rather than concentrating on content. Of course, in the end you must attend to both form and content if you want to communicate effectively with your readers.

Many word processing programs offer formatting features such as underlining, boldface, italics, and so on, with which you can vary the appearance of your text and highlight important information. Be certain, however, that you check with your instructor to ascertain his or her preferences for format style before varying the style too much. Your main goal should be to make your paper professional in appearance. Thus, an English Gothic typeface that prints in scrolling capital letters, for example, would not be appropriate for a research paper. Nor are margins that are justified (even) on the right of the page typically appropriate for a formal paper. It is best to be conservative and justify left margins only.

Proofreading

Once your paper has been typed to your teacher's specifications and you have run your word processing program's spell-check feature, you will still need to proofread carefully for any errors not caught by the computer. For example, spell-check cannot tell you if you have used "their" when you should have used "there." It is best to print out a clean copy of your paper after making all of your proofreading corrections. However, your teacher may not object to your making a few minor corrections on the paper, preferring that you correct any errors, even though this may necessitate some handwriting on the typed page, rather than leaving them uncorrected.

One helpful way to proofread for typing errors is to begin at the bottom of the page and read up one line at a time. In this way, you keep yourself from reading for meaning and look only at the form of the words. You can spot errors more easily when you are not actually reading the paper. If you are proofreading on a computer screen, you can use your search command to search for periods from the bottom of the text upwards. In this way, your computer's cursor will skip to the previous sentence, thus reminding you to read it independently.

Keep your dictionary handy and refer to it whenever you have any doubt about the appropriate use of a word. Use your grammar and usage handbook to double-check any last minute questions about grammar and punctuation. If you have access to grammar-check software, such as *Writer's Helper, Correct Grammar,* or *Grammatik,* it is sometimes helpful to run it on your paper, keeping in mind the cautions about such programs that were discussed previously.

It is impossible to overstress the importance of careful proofreading by you. Even if the paper was word processed or typed by a professional typist, you will probably find errors when proofreading. Since you are the paper's author, any errors are your responsibility, not the computer's or the typist's. It is a good idea to save early drafts of your paper even after the paper has been typed. Early drafts serve as a record of your thinking and your work on the paper. If you have taken care at every stage of the revision process, your paper will be one you can be justifiably proud of.

CONSIDERING FORMAL DETAILS

The formal details outlined in the paragraphs that follow incorporate some general principles in research writing. However, it would be wise to also consult the relevant style manual for your field to discover any minor variations from this format.

Always type research papers or have them typed for you. Use a standard typeface or font and a fresh printer ribbon. It might be wise to ascertain whether or not your teacher prefers that papers not be printed on dot matrix printers. Sometimes these are light and difficult to read. If possible, print your final copy on a laser printer for a professional appearance and ease of reading. The paper should be a standard weight and size (8½ × 11 inches). If you do not have a self-correcting typewriter or a word processor, use Liquid Paper or correction tape to correct errors.

Spacing

Use only one side of the paper and double space all the way through, even for long quotes that are indented in the text (Note: Styles vary). Leave four blank lines between major sections, three or four between heading and section, and three or four above and below indented, long quotes (more than four lines of text). Also, double space the endnote page (if used) and references page. Some journals ask writers to submit their papers with each section beginning on a separate sheet of paper (e.g., Abstract, Introduction). This is usually not necessary for a student paper, but do put any figures or tables on separate pages, numbered consecutively (one set of numbers for figures; another for tables). Place these illustration pages after the Works Cited page for reference.

Margins

Use a margin 1 to 1.5 inches wide on all sides of each sheet. Your word processing program allows you to set the margins appropriately. If you are typing your paper, use a typing guide (a sheet of paper that goes behind the sheet you are typing on and whose dark ruled lines show through), set the margins on your typewriter, or mark each sheet with a pencil dot one inch from the bottom so you will know when to stop typing on a page. Sophisticated word processing programs will help you format the pages appropriately. In particular, if you are using footnotes, these can be generated automatically by some word processing programs, which will leave the appropriate amount of space for footnotes on each page. If you are typing footnotes at the bottom of the page yourself, plan your bottom margins very carefully to allow room for the notes. Generally, endnotes are easier to type and perfectly acceptable in most cases.

Title

Ask your instructor whether you need a title page. If the answer is yes, find out what information should appear there. Generally, title pages contain three kinds of identifying information: the title of the paper, author identification, and course identification (including date).

If you do not need a separate title page, put your name, date, assignment name, and any other identifying information on the upper right-hand corner of the first page. Center the title on the first page three or four lines below the identifying information or, if you use a separate title page, one inch from the top of the page. The title should not be underlined, surrounded by quotation marks, or typed in capital letters. Leave three or four lines between your title and the beginning of the text.

Numbering

Number each page starting with the first page of text after the title page. (Note: some styles omit the page number from page 1.) Place the numbers in the upper right-hand corners or centered at the bottom of the pages. Your word processing program allows you to automatically number your pages and to suppress numbering on any pages where they are not needed. For example, you typically don't need to number the endnote page and the references page. Rather, identify them with the appropriate heading centered one inch from the top of the page and followed by three or four blank lines. Some styles recommend headers along with the page numbers (for example, Hult - 2). Check your word processing manual for help in generating such automatic headers.

Indentation

Use uniform indentation for all paragraphs (five spaces is standard). Indent long quotes (more than four lines long) five spaces from both right and left margins or ten spaces from the left margin only. Indent the second and subsequent lines of the reference-list entries five spaces. Leave two spaces between each sentence and after a colon or semicolon. Divide words at the end of lines according to standard rules. Use your dictionary if you are unsure of where to divide a word.

The Abstract

An abstract is a very short summary of a paper, usually one tenth to one twentieth the length of the whole. The purpose of an abstract is to condense the paper into a few, succinct lines. Thus, the reader must be able to understand the essence of the paper from reading just the abstract, without actually reading the paper. Your abstract should cover the purpose of your paper as well as the major topics you discuss. To write an abstract, follow the same general procedure you used to write a summary paper. However, you will need to compress information into a few compact sentences. Even though the information in your abstract is necessarily densely packed, it should still be readable and understandable.

The Endnote Page

If your paper will have endnotes, type them on a separate page immediately after the text of your paper (and before the references page). Center the title, "Notes" or "Endnotes," one inch from the top of the page, and type it in capital and lower-case letters (not all capitals). Do not use quotation marks or underlining. Leave three or four blank lines between the title and the first line of your notes. Type the notes in consecutive order based on their appearance in the text. Indent the first line of the note five spaces from the left margin, type the superscript number, and leave a space before beginning the note. For any run-over lines of each note, return to the left margin.

The References Page

Center the title "References," "Works Cited," or "Bibliography" and type it one inch from the top of the page in capital and lower-case letters (not all capitals). Do not use quotation marks or underlining. Leave three or four blank lines between the title and the first line of your references. The references themselves should be typed, double spaced, and listed in alphabetical order by the author's last name (or the title, if the author is not known). (Note: In the number system, references are listed consecutively as they appear in the text.) To make the alphabetical list, sort the bibliography cards (on which you have recorded the sources actually used in your research paper) into alphabetical order and transcribe the information in the proper form from the cards to your list. The references page follows the last page of your paper or the endnote page (if included) and need not be numbered.

The Annotated Bibliography

In some cases it is helpful to provide your readers with more information about the sources you used in your research than is typically given in a bibliography. An annotated bibliography serves this purpose. To construct an annotated bibliography, you would first compile all of your references, alphabetize them, and format them according to the documentation style for your discipline. Then, following each bibliographical entry, you would state in a sentence or two the gist of the source you had read and its relevance to your paper. An annotated bibliography can help your readers to decide which of your sources they would like to read themselves. It should not be difficult for you to annotate (that is, provide brief glosses) for sources that you have used to write your paper.

The Appendix

Material that may not be appropriate to the body of your paper may be included in an appendix. You may use the appendix for collations of raw data, descriptions of primary research instruments, detailed instructions, and so on. The appendix is located after the bibliography or references page and is clearly labeled. If there is more than one appendix, label them Appendix A, Appendix B, and so on. When referring to the appendix in the paper itself, do so in parentheses: (For a detailed description of the questionnaire, see Appendix A.)

Using Graphics

Sometimes you will find that you need to illustrate your work using graphics. Types of graphics include tables (usually showing quantitative data), figures (charts, graphs, or technical drawings) or illustrations (including photographs). You first need to determine whether or not your readers would benefit from the inclusion of a graphic. Sometimes repeating information in another form, such as a graphic, will help readers to remember it. The following principles may help you decide:

Use a graphic when words alone cannot describe a concept or object adequately;

Use a graphic to summarize an important point;

Use a graphic when it can help to conveniently display complex information.

You also should determine purposes for the graphic to accomplish and where it would best be included in your paper. To align your graphic, remember that the graphic should be easy to read and displayed with ample white space so that it stands out from the text background. You should refer at least once to each graphic, either directly or parenthetically, in the text itself. The graphic should be placed on the same page in which it is mentioned and labeled with both a title and a source. Number tables sequentially throughout your paper; if you are including both figures and tables, they should be numbered separately (e.g., Table 1, Table 2, Figure 1, Figure 2]. (For examples of tables and figures, please see the sample papers in Chapter 4.)

6

Secondary Research Methods: Writing a Review Paper

As discussed in Chapter 1, sciences in general attempt to explain phenomena in the natural and physical world. Since scientists rely on current technology as tools to help them in their work, technology itself has become a branch of science. Scientific researchers must have knowledge of the current research being conducted by others in their fields. They must also have knowledge of the technologies needed to conduct that research. Although the experimental method is at the heart of scientific research, library research is also important. It is in the scientific journals and reviews that scientists report their findings for scrutiny and replication by other researchers.

The second of the two major types of papers typically written in science and technology is discussed here: the review paper. In the scientific review, sometimes called a literature review, you analyze for your readers the present state of knowledge in an ongoing field of research as it has been reported in books and journals. Your contribution, then, is in the way you interpret, organize, and present the complex information, thus making it easily accessible to the reader. In the scientific review paper, you will argue a particular position with support from the literature and review the topic through paraphrase and summary as objectively as you can, but you won't introduce any new primary data. Several library research principles and skills are important for you as you investigate the topic you have chosen to review. These include:

1. A familiarity with library research tools, including databases, bibliographies, and indexes used in science and technology.
2. The ability to understand and evaluate information and data from a variety of sources.
3. The ability to paraphrase and summarize information and data in your own words.
4. The ability to synthesize the information and data gathered into an organized presentation.
5. The ability to employ the formal conventions of scientific review papers.

A GUIDE TO THE SCIENTIFIC REVIEW PAPER

To begin your review paper, first determine a topic and narrow it to a manageable, researchable size. If you are taking a science course now, your textbook is a good place to begin looking for research ideas. Check the table of contents in your textbook and in any references that may be listed. Another source of ideas is current scientific journals. Sheldon Cooper, whose research serves as the model for this chapter, became interested in the subject of bat muscles when he read about bats in a biology class. Be sure you select a topic that will hold your interest and attention, preferably a topic that you already know something about, so that you will be an informed and objective reviewer.

Preparation

Gather the necessary materials for your research project: a notebook or notecards. Sheldon decided to use a file on a computer disk as a computerized research notebook (see p. 35 for a discussion of using a computer in this way). Make a schedule that gives you several weeks to conduct your library research and several weeks to write a draft and final copy of your review paper. If you foresee that your research project will contain primary research data, make sure you allow yourself sufficient time to gather that data. Careful planning at the outset of such a major project ensures that you have enough time to carry out the research necessary to write an informed review.

Developing a Search Strategy

Sheldon had formulated some general impressions about bat muscles through reading for his classes. He wanted to discover whether bat muscle composition progressed through seasonal variations. His library search began with a look at general background sources in life sciences and moved to more specific works on bat muscles (your particular search may differ somewhat from this outline):

Sheldon's Search Strategy

1. Look up bats in background sources from the reference area of the library, including the *Henderson's Dictionary of Biological Terms*, 10th ed.
2. Look up reviews and articles on the subject using *Current Contents: Life Sciences*.
3. Use the library online catalog to find books and documents on bat muscles; be sure to check appropriate terminology in the Library of Congress Subject Headings.
4. Use database indexes, print indexes, and abstracts for access to scientific journal articles: the *Biological Index*, the *Applied Science and Technology Index*. Search by subject headings from LCSH and by key words.

The Library Reference Area

Many students begin their research at the library catalog; however, you may find it more profitable to begin with the general sources in the reference area of your library. In a scientific review, it is usually important to obtain the most current library information. Since books take years to write and sometimes years to produce, even the most recent editions of books can contain information that is three or four years old. The most recent information is probably in journals, which generally publish papers one or two years after the studies are conducted and the paper written.

The reference area of your library provides you with valuable background and contextualizing information on your topic. Plus, the reference area typically contains the tools, such as indexes and computers, that give you access to a variety of other materials. Reference materials are usually arranged by Library of Congress call numbers, which are grouped by subject area. If you cannot easily discover on your own how your reference area is organized, do not hesitate to ask your reference librarian for help.

To illustrate the use of a library search strategy in the sciences and technology, we will follow the search steps above using Sheldon's research on bat muscles as a model for your own search.

◆ **EXERCISE**

Outline your own search strategy, beginning with general and working to specific sources. Draw up a research time schedule.

General Sources

By reading general information about your topic, you can put it into a context and start focusing your search—narrowing your topic to a manageable size. Also, in reading general and specialized encyclopedias and dictionaries, you will learn what is considered common knowledge on the topic. Depending on your particular topic, you may be reading general encyclopedias, such as *Britannica*, and specialized encyclopedias, such as the *McGraw-Hill Encyclopedia of Science and Technology*. Specialized sources for science and technology are on pages 26–31. Sheldon used the following background sources, which he listed in his computerized research notebook at the beginning of his working bibliography:

Henderson's dictionary of biological terms, 10th ed.
   Eleanor Lawrence, ed. New York: John Wiley, 1989.
Encyclopedia americana. International edition. New
   York: Grolier, 1990.

At the end of each source, such as your textbooks or the encyclopedia previously mentioned, you usually find a list of bibliographic citations and references. This reference list can be an important place to locate key sources and reports written about your topic. List any promising references on your working bibliography, because later they may lead you to valuable information. It is not necessary for you to look up each entry on your working bibliography at this time.

Focusing Your Search

After you have read several background sources about your topic, you are ready to narrow the subject of your review. Sheldon, for example, decided to report on the current state of knowledge regarding seasonal variations in bat muscles. Sheldon developed the following starting questions to guide his research: "During hibernation, do the histochemical and biochemical properties of bat muscles change? Are there differences in the composition of bat muscles during hibernation as compared to during summer?" These starting questions provided Sheldon with a direction for his research.

✦ EXERCISE

Define for yourself just what topic you are trying to review and what specifically within that topic you will cover in your review paper. Then write a starting question that you intend to investigate in your research.

Reviews

For many scientific review papers, reviews and reports of research that have already been written on the topic are useful. Scientists customarily publish reviews of current scientific studies to help other scientists keep abreast of the field. Even though you are writing a scientific review yourself, it might be helpful to see what others have to say on the subject before you begin. You may need to update somewhat any review you find by reading the most current sources in the field.

Reviews of research and research reports can also provide you with reference lists from which to build your working bibliography. Generally, scientific disciplines publish annual reviews in a particular journal within the field. In the library online catalog, look up the field, for example, chemistry, and find the annual review journal. If your library carries the *Index to Scientific Reviews,* you can locate relevant reviews there as well. If you have difficulty finding reviews from a particular field, do not hesitate to ask your reference librarian for assistance.

Another useful tool for finding current articles and reviews is the *Current Contents* journal, which indexes articles and reviews by discipline. The following are some examples of *Current Contents* journals relevant in the sciences:

Current Contents: Life Sciences
Current Contents: Physical, Chemical, Earth Sciences
Current Contents: Engineering and Technology
Current Contents: Agriculture, Biology, and Environmental Sciences

The *Current Contents* journals are particularly useful because they allow easy and quick access to current articles and reviews written on a particular subject. *Current Contents* publish weekly issues, which include the tables of contents from the latest journals in a particular discipline. The journals are indexed by key title words and by authors. In our library, the *Current Contents* journals are provided in a computerized database; they are also available in print. By typing the keywords <u>bats and muscles</u> into the computer while searching the database for *Current Contents: Life Sciences,* Sheldon discovered the following recent article, which he listed in his working bibliography:

```
Brigham, R. M., C. D. Ianuzzo, and T. H. Kunz. 1977.
    Histochemical and biochemical properties of
```

```
flight muscle fibers in the little brown bat,
Myotis lucifugus. J. Comp. Physiol. 119:141-154
```

✦ EXERCISE

Locate and read reviews of research available on your topic. See the end of this chapter for a list of sources and databases used in the sciences and technology.

The Library Catalog

At this point in your search you will want to use the library online catalog. First, use the catalog to help you track down the book sources you may have already listed on your working bibliography from background sources. Once in the database that lists your library's book holdings, you will indicate that you want the computer to search by title (e.g., t = animal physiology). If this book is available in your library, it will be listed, along with its call number to help you locate the book.

The most important and powerful use for the online catalog is its cross-referencing function: using the subject headings and keywords to locate additional titles on the same or related topics. You want to instruct the computer to search for books on your subject (for example, s = bats—physiology) or by keyword or combinations of keywords (for example, k = bats and muscles). (For more information on using the online catalog, see Chapter 2.)

Sheldon used the LCSH list to discover headings under which his topic might be listed. He found the following headings to be useful: Bats, Bats—physiology, Bats—flight. Using these headings in the online catalog, Sheldon found the following titles:

```
Eckert, R; Randall, D.; and Augustine, G. 1988.
    Animal physiology: Mechanisms and adapta-
    tions. 3rd ed. W. H. New York: Freeman and
    Company.
Hainsworth, F. R. 1981. Animal physiology: Adaptations
    in function. Reading, MA: Addison-Wesley.
Schmidt-Nielsen, K. 1990. Animal physiology:
    Adaptation and environment. 4th ed. Cambridge
    Univ. Press.
```

◆ **EXERCISE**

Use the library online catalog's subject and keyword searching to find additional books and materials on your topic. Use the LCSH list to find the appropriate subject headings under which sources for your topic are listed.

Indexes and Abstracts

Once you have gathered a substantial amount of information on your subject and have located several key references and books, you are in a position to expand your bibliography by gathering additional articles found in professional journals and periodicals. Helping you find articles on your subject is the principal function of subject indexes, citation indexes, and abstracts. These referencing tools are generally found either online or in print.

Computerized Indexes and Abstracts

Most modern libraries rely heavily on computer databases to help patrons search for professional journal articles. Our library, for example, contains databases of all the indexes produced by the W. W. Wilson company; these databases can be searched by individuals via computer. Our screen initially shows a menu of possible databases to search: one for general periodicals; one for professional literature in the sciences, agriculture, and engineering; one for professional literature in the arts and humanities; and one for professional literature in the social sciences. After selecting the appropriate database for your topic, you can search the database using the subject headings you found while searching for books, or you can use keywords and combinations of keywords. Sheldon used the key words "k=bats and muscles" to locate articles in the science journal database of the library's online catalog. This combination of terms yielded exactly the articles that he was looking for in his search, including the following:

Hermanson, J. W., LaFramboise, W. A., Daood, M. J.

    Uniform myosin isoforms in the flight muscles of

    little brown bats, <u>Myotis lucifugus</u>. The Journal

    of Experimental Zoology 259:174-80 Aug '91

Other databases may be available in your library through computers that make use of compact disks (CD-ROMs). These databases usually index specialized professional articles. Our library, for example, currently offers patrons CD-ROM databases to search for articles in the areas of business, agriculture, education, medicine, psychology, natural resources, and wildlife. Check with your library to find out whether any indexes via computer are available to you. Searching in this way is far more efficient and thorough than searching through print media. (NOTE: Many of these databases are indexed using "controlled vocabularies" other than the LCSH. Be sure to check for a "thesaurus" or listing of subject headings for each database you use.)

Print Indexes

The subject indexes, such as *General Science Index* or *Applied Science and Technology Index,* list articles published in a given year by subject and author. By using these indexes, you can search for citations to journal articles written on your topic. You should usually start with the most recent volume of the index and work your way back, looking up your topic in several volumes of the index.

The second type of print index you may need to use in your library search is the citation index. Through the use of the citation index, you can begin with a particular researcher's name and work your way forward to other researchers who have listed (cited) that researcher in their subsequent work. Citation indexes are relatively comprehensive listings of such citations. The key sources listed in other reference works are the cited sources in the citation index. Typically, you will know the names of key researchers on your topic after a thorough search of the encyclopedias and reviews of research. Then you can follow up by searching each volume of the citation index for citations to these key sources that have appeared since the original publication of the key source. In this fashion, you will quickly build your working bibliography.

The most important citation index for the sciences is the *Science Citation Index.* Sheldon, for example, knew that R. Eckert was a major researcher on animal physiology, because he had written several textbooks on the subject. Sheldon looked up the name of the key source, R. Eckert, in the most recent citation volume of the *Science Citation Index.* He found Eckert's 1988, *Animal Physiology,* listed as a cited source (meaning that authors had used his work in subsequent research). Under Eckert's name were listed the researchers who had used his work as a basis for their own (that is, the citing sources). Sheldon added the following source to his working bibliography as found in the citation index:

```
Vanderwa, JG Anim Produc 50 277 90
```

Listed first is the name of the author who cited Eckert's work; then comes the abbreviated title of the journal, followed by publication data—volume number, page number, year. To find the full title of the journal, you must look at the abbreviations list at the front of the source volume. In this case, the abbreviation refers to the *Animal Production Journal*.

It may take you some time to become familiar with how the citation indexes work, but doing so will be well worth the effort. These indexes are a major tool in the sciences. When using the citation index, you may find that the complete title of the citing source is omitted, so it may be difficult to know whether the article will turn out to be relevant to your search or not. By looking up the author's name in the source index for the same year, you may find a more complete listing. You should review each article later to determine its importance to your search.

A third important type of indexing service is abstracts. Sheldon found *Biological Abstracts* to be useful in his search. Abstracts go one step further than indexes by providing you with a short summary of the article, which can help you sort through your references for those that are the most appropriate for your own search. Some abstracts may be available on computers: *Biological Abstracts*, for example, is available in a CD-ROM database. Check with your librarian to find out which abstracts in your library may be searchable via computer. (Note: Even when using abstracts, it is important to be careful not to plagiarize. If you use information from an abstract, it needs to be cited. See page 148 for information on citing electronic abstracts.)

✦ EXERCISE

Use subject and citation indexes to find titles of articles related to your topic, using both subject headings and keywords to search. Do not limit your search to print media only; also search any relevant databases by subject and keyword. For a complete list of indexes and abstracts in the sciences and technology, see pages 29–31.

Evaluation

Once you have located a book or an article, immediately evaluate it for its relevance or usefulness in your search. It is not unusual for a book or article with a very promising title to turn out to be something totally different from what you expected to find. Or you may discover a controversy in the field that you were not aware of prior to your search. As you review your sources, continually sort through and discard any that are not relevant. If, after an initial screening, a book looks as though it could be useful to you, check the book out at the circulation desk. In the case of articles, either photocopy them for later use or take notes from them in the library, since they are generally noncirculating materials and cannot be checked out.

◆ EXERCISE

In your research notebook, evaluate each article and book to be used in your research paper. Follow the source evaluation guidelines in Chapter 3 on page 55.

Taking Notes

As you begin to take notes on the sources, remember to record complete bibliographic information so that you will not need to look up a particular source again. If you are using a research notebook (whether on paper or in a computer file), complete information must be kept on each source you use in your research, including author(s), title, and publication data. Taking care at this stage will benefit you when you get to the actual writing stage.

If you are taking notes in your research notebook or in a computer file, take care to identify each source as you are writing down your notes. Put into quotation marks any information or wording taken directly from the source, and at the end of the source information, mark down the exact page number on which you found the material and whether you paraphrased or quoted the author.

ORGANIZING AND WRITING THE SCIENTIFIC REVIEW PAPER

A major task in writing a scientific review is organizing the material you have gathered. It is your job to make sense of the information you found in your library search. Remember, you are trying to make the information accessible to your readers as well as objective and comprehensive. When Sheldon began to narrow his topic, he decided to focus on the seasonal variation issue with bat muscles. As he thought through the rhetorical situation, he decided that he wanted to convey to an expert audience just what the current state of knowledge was regarding seasonal variations in bat muscles.

After completing his library research, Sheldon articulated an answer to his starting question in the form of a thesis statement: "This paper will examine changes in histochemical and biochemical properties of bat muscle during hibernation compared with summer." This thesis statement provided Sheldon with an overall direction for his paper.

To understand the seasonal variations, Sheldon decided to first describe the types of skeletal muscle fibers found in mammals; then he would discuss the histochemical variations in bat muscle, followed by the biochemical variations in bat muscle. This organizational plan made the information easily available to the reader. He divided each subsection with descriptive headings to further help the reader discern his organizational plan. Scientific reviews often are subdivided in this way to allow the reader easy access to the information.

✦ EXERCISE

Write a thesis statement and sketch a preliminary organizational plan for your research paper.

Arranging the Materials

Once Sheldon had decided on the thesis statement and organizational plan, he grouped related information in his computerized research notebook, using the cut and paste functions of his word processor, under the headings he had decided on in planning his paper. Because all his information was stored in his own words in his computer research notebook, it was an easy matter to move blocks of material in the notebook using the block move (cut and paste) commands; furthermore, he didn't need to worry about plagiarizing any of the sources because he had already been very careful to summarize and paraphrase while taking notes. (Be sure to make a backup file of your notebook before beginning to manipulate the information in this way.)

Sheldon moved to the end of the computer file any information that did not seem to fit into the paper, such as information he had gathered on the specific diseases. This material was not relevant to his particular thesis statement. To make a unified, coherent presentation of your research, you must discard any information that is irrelevant. With his preliminary plan set, Sheldon wrote a more detailed outline to guide him in writing the paper:

> Outline
> I. Introduction
> II. Mammalian skeletal muscle fiber types
> III. Histochemical variation in bat muscle
>   A. Pectoralis muscle

B. Accessory flight muscles
  C. Gastrocnemius muscle
 IV. Biochemical variation in bat muscle
  A. Pectoralis muscle
  B. Gastrocnemius muscle
  C. Cardiac muscle
 V. Summary

✦ EXERCISE

Sort your notecards by their titles, number related ideas in your research notebook, or block related information together in a computer file. Write an informal outline of your research paper, using your thesis statement and organizational plan as a guide. Begin writing the first draft of your research paper.

Writing the First Draft: Verification

After you have completed your outline, you are ready to write the first draft of your research paper. Remember, you are writing in order to review for your readers the current state of thinking on a particular scientific topic. Remind yourself at this time of the general understanding you had of your topic and of your answer to the question you posed as you began researching. When writing your first draft, use concrete and simple language to explain as objectively as you can the current thinking on your topic. Your outline will guide the writing of this first draft. Any word, phrase, or sentence you copy directly from a source must be placed in quotation marks, followed by the last name of the author, the date the source was published, and the page number (in parentheses):

> The body's T cells "are responsible for the ability of vertebrate animals to recognize that antigens, or foreign materials, have invaded their bodies" (Marrack and Kappler 1988, p. 36).

(Note: In the number system, sources are identified by a superscript number or a number in parentheses immediately following the sources instead of the author's name, date, and page number in parentheses. See The Number System later in this chapter.)

Similarly, paraphrases and restatements of ideas taken from a source should be given documentation even though you have recast them in your own words:

```
Microbes can easily enter the blood from the
lymphatics and spread infection throughout the
body (Wyss 1971).
```

Remember: you do not need to document common knowledge on your topic.

For general information on planning, writing, and revising your scientific review paper, refer to Chapter 5. Use the following information on documentation in the sciences and technology to cite sources in the correct form. The sample review paper at the end of this chapter serves as a model of a scientific review conducted on a limited, accessible topic.

DOCUMENTATION IN SCIENCE AND TECHNOLOGY

There is no uniform system of citation in the sciences and technology, but all disciplines follow either a journal style or a style guide; therefore, some general principles apply to most scientific disciplines. The sciences use in-text citation and list the works cited at the end of the text. The recency of the source is important, so the year of publication is stressed in the citation. Entire journal articles rather than specific pages may be cited, and direct quotes are seldom used.

Internal Citation

In the sciences, authors are cited within the text itself by means of either the author/year system or the number system.

The Author/Year System

The author/year system is widely used in the sciences and has been adopted (with variations) by the social sciences and business. It is a fairly easy system for the reader to use. The following principles should be observed:

1. When an author's journal article in general is cited, the source material is followed by the last name of the author and the date of the source article in parentheses:

```
The T cell plays a key role in immunology
(Marrack and Kappler 1988).
```

2. If the source material is paraphrased or directly quoted, the page numbers should be included:

   ```
   Analysis indicates that "binding does not change
   the structure of interacting units" (Davis 1987,
   p. 134).
   ```

3. If the author's name is used in introducing the source material, only the date is necessary:

   ```
   According to Davis (1987) scientists are on the
   verge of discovering the secrets of antibodies.
   ```

4. Multiple sources may be cited:

   ```
   Recent research indicates that antibodies may
   also bind to microbes and prevent their
   attachment to epithelial surfaces (Bellanti 1985;
   Getzoff 1987; Geyson 1987).
   ```

5. When citing a work with multiple authors, use et al. ("and others"):

   ```
   The earlier work suggested other substance
   involvement in fish fertilization (Henkert et
   al. 1978).
   ```

6. When citing a work with two authors, join them with *and*:

   ```
   Similar observations have been made in sea
   urchins (Shumomura and Johnson 1976).
   ```

7. Information obtained from another work cited within the first work should appear as follows:

   ```
   Such a factor is apparently not present in
   unactivated sturgeon (Chulitskai 1977, cited by
   Meyerhof and Masui 1980).
   ```

8. When the same author has written two or more publications in the same year, designate them with an *a, b,* and so on following the year:

```
Some of the earliest responses of eggs to
sperm-egg interaction are electrical (Nuccitelli
1980a).
```

The Number System

The number system is also used in the sciences and technology. Here a number is assigned to each source listed on the references page. To cite the source within the text, one simply lists the number of that source, either in parentheses or as a raised superscript:

```
Temperature plays a major role in the rate of
gastric juice secretion (3).
```

```
Temperature plays a major role in the rate of
gastric juice secretion.³
```

You can cite multiple sources easily with this system:

```
Recent studies (3,5,8) show that antibodies may
also bind to microbes and prevent their
attachment to epithelial surfaces.
```

Recent studies[3,5,8] show that antibodies may also bind to microbes to prevent their attachment to epithelial surfaces.

Content Notes

Some scientific papers might require notes that explain something about the text itself rather than refer to a particular source being cited. These content notes are listed either as footnotes or endnotes rather than internal citations.

The Reference List

The reference list, found at the end of your research paper, contains all the sources actually used in the paper. The title of this page is "References," "Works Cited," or "Literature Cited." The purpose of the reference list is to help readers find the materials you used in writing your paper. Therefore, you must give complete, accurate information here. The following principles are generally accepted for the reference list in the sciences and technology:

1. On the references or work-cited page, references are arranged in alphabetical order and may be numbered. (Note: The numbering system may proceed consecutively, i.e., in the order in which the sources appear in the text.)
2. Authors are listed by surnames and initials.
3. Generally the first word only of a title is capitalized, the title of an article is not enclosed in quotation marks, and the title of a book is not underlined.
4. Names of journals are often abbreviated.
5. The volume and page number system often resembles that found in the indexes (for example, 19: 330–360). Sometimes the volume number is in boldface type, indicated by a wavy line in manuscript: **16,** or 16.
6. The year of publication appears either immediately after the author's name or at the close of an entry, depending on the particular journal's publication style.

BOOK

Golub, E. S. 1987. Immunology: A synthesis. Boston: Sinaur Associates.

ARTICLE

Milleen, J. K. 1986. Verifying security. ACM Computing Surveys 16:350-354.

or

Milleen, J. K. Verifying security. ACM Computing Surveys 16:350-354; 1986.

7. If the same author has published two or more works in the same year, indicate this with a lowercase a and b: 1984a, 1984b.

8. *Author/year system:* The first word of the entry is typed at the left margin. Subsequent lines of the same entry are indented five spaces. Generally, the entire reference list is double spaced. *Number system:* The numbers are typed at the left margin. The first line of each entry is typed two spaces after the number. Subsequent lines are even with the first line.

If you are writing a paper for a specific discipline, it is important for you to find out which documentation form your instructor prefers. Some style guides that will help you:

The ACS Style Guide: A Manual for Authors and Editors. Janet S. Dodds, ed. Washington, D.C.: American Chemical Society, 1986.

Council of Biology Editors Style Manual. 5th ed. Council of Biology Editors, 1983.

Geowriting: A Guide to Writing, Editing, and Printing in Earth Science. 3rd ed. American Geological Institute, 1979.

Style Manual: For Guidance in the Preparation of Papers for Journals Published by the American Institute of Physics. 3rd ed. New York: American Institute of Physics, 1978.

(NOTE: some scientific journals follow the APA style as outlined by the *Publications Manual of the American Psychological Association.*)

The model references in the accompanying table are based on the form used in many science journals. They follow the style found in the *Council of Biology Editors Style Manual* (CBE). (A few of the model references are taken directly from CBE.) For further examples, refer to one of the manuals listed in the previous paragraph. (Please note the position of the date in CBE style. Alternatively, many science journals place the date immediately after the author's name.)

MODEL REFERENCES: NATURAL AND PHYSICAL SCIENCE (CBE)

Type of Reference

BOOKS

1. One author

```
Campbell, R. C. Statistics for biologists. 2d ed.
    London and New York: Cambridge Univ. Press;
    1974.
```

2. Two or more authors

   Snedecor, G. W.; Cochran, W. G. Statistical methods. 6th ed. Ames, IA: The Iowa State Univ. Press; 1967.

3. Two or more books by the same author (list chronologically, or use a and b if published in the same year)

   Parker, D. B. Crime and computer security. Encyclopedia of computer engineering. New York: Van Nostrand Reinhold; 1983a.

   Parker, D. B. Fighting computer crime. New York: Scribner's; 1983b.

4. Book with an editor

   Buchanan, R. E.; Gibbons, N. E., editors. Bergey's manual of determinative bacteriology. 8th ed. Baltimore: Williams and Wilkins; 1974.

5. Section, selected pages, or a chapter in a book

   Jones, J. B.; Beck, J. F. Asymmetrical syntheses and resolutions using enzymes. Jones, J. B.; Sih, C. J.; Perlman, D. eds. Applications of biochemical systems in organic chemistry. New York: Wiley; 1976: pp. 107-401.

6. Book with a corporate author

   American Society for Testing and Materials. Standard for metric practice, ANSI/ASTME 370-376. Philadelphia: American Society for Testing and Materials; 1976.

7. Work known by title

American men and women of science. 13th ed.
  Jacques Cattell Press, ed. New York: Bowker;
  1976. 6 vol.

8. All volumes in a multivolume work

Colowick, S. P.; Kaplan, N. O. Methods in
  enzymology. New York: Academic Press;
  1955-1963. 6 vol.

ARTICLES

1. Journal article (one author)

Solokov, R. Endangered pisces: The Great Lakes
  whitefish is exploited by both lampreys and
  humans. Nat. Hist. 90:92-96; 1981.

2. Journal article (two or more authors)

Berry, D. J.; Chang, T. Y. Further
  characterization of a Chinese hamster ovary
  cell mutant defective. Biochemistry
  21:573-580; 1982.

3. Article on discontinuous pages

Balack, J. A; Dobbins, W. O. III. Maldigestion
  and malabsorption: Making up for lost
  nutrients. Geriatrics 29:157-160, 163-167;
  1974.

4. Article with no identified author

Anonymous. Frustrated hamsters run on their
  wheels. N. Sci. 91:407; 1981.

[Note: May also be listed by title only]

5. Newspaper article (signed)

>Shaffer, R. A. Advances in chemistry are starting to unlock the mysteries in the brain. The Wall Street Journal. 1977 Aug. 12; 1 (col. 1), 10 (col. 1).

6. Newspaper article (unsigned)

>Puffin, a rare seabird returns to where many were killed. The New York Times. 1977 Sept. 6; Sect. C: 28.

7. Magazine article

>Starr, D. Students who tap the Universe. Omni. 1989 May; 66-72.

TECHNICAL REPORTS

1. Individual author

>Brill R. C. The TAXIR primer. Occasional paper—Institute of Arctic and Alpine Research. 1971; 71p. Available from: Univ. of Colorado, Boulder, CO.

2. Corporate author

>World Health Organization. WHO Expert Committee on Filariasis: 3d report. WHO Tech Rep. Ser 542; 1974. 54p.

3. Government document

>U.S. Congress, House of Representatives. The international narcotics control community. A report on the 27th session of the U. N. Commission on Narcotics to the Select Committee on Narcotics Abuse and Control.

Ninety-fifth Congress, first session. 1977 Feb. 37p. Available from: U.S. Government Printing Office. Washington, DC: SCNAC-95-1-10.

OTHER SOURCES

1. Motion picture

   Rapid frozen section techniques [Motion Picture]. U.S. Public Health Service. Washington DC: National Medical Audiovisual Center and National Audiovisual Center; 1966. 6 min.; sd; color; super 8 mm; loop film in cartridge; magnetic sound track.

2. Dissertation or thesis (unpublished)

   Dotson, R. D. Transients in a cochlear model. Stanford, CA: Stanford Univ.; 1974. 219p. Dissertation.

3. Letters and interviews

   Darwin, C. [Letters to Sir J. Hooker]. Located at: Archives, Royal Botanical Gardens, Kew, England.

   Quarnberg, T. [Interview with Dr. Andy Anderson, Professor of Biology, Utah State University]. 1988 April 25.

4. Unpublished paper presented at conference

   Lewis, F. M.; Ablow, C. M. Pyrogas from biomass. Paper presented to Conference on capturing the sun through bioconversion. Washington, DC; 1976. Available from: Stanford Research Institute, Menlo Park, CA.

5. Reference work

   Handbook of psychopharmacology. Section I: basic neuropharmacology. Iverson, L. L.; Iverson, S. D.; Snyder, S. H., eds. New York: Plenum Press; 1975. 6 vol.

6. Abstract on CD-ROM

   Rodriguez, A. M. 1991. Multicultural education: Some considerations for a university setting [CD-ROM]. Abstract from: SilverPlatter's ERIC Item: ED337094.

7. Online abstract

   Lawrence, O. J. 1984. Pitfalls in electronic writing land. [Online]. English Education, 16.2; 94-100. Abstract from: Dialog file: ERIC Item: EJ297923.

8. Online journal article

   Herz, J. C. 1995, April. Surfing on the internet: A nethead's adventures online. [Online serial]. Urban Desires, 1.3. Available Internet: www/desires.com/ud.html.

9. Electronic correspondence, such as e-mail messages and conversations via bulletin boards and electronic discussion groups, is typically cited as personal communication in the text. In-text information to include:

   author, date, subject of the message
   name of the listserv, bulletin board, or e-mail discussion group
   available from: e-mail address

   (For further information on citing electronic information, see Li, X.; Crane, N. B. Electronic style: A guide to citing electronic information. Westport and London: Meckler; 1993.)

EXERCISES AND RESEARCH PROJECT

Complete the exercises outlined in this chapter as you research a limited scientific or technological topic and write a scientific review paper. The three exercises that follow will give you additional practice using skills associated with science research projects.

1. For each entry on your reference list, write a three- or four-sentence annotation that describes the content of that source.

2. Write a "review of the literature" report that summarizes in three to four pages the major ideas found in your sources. In your review, try to avoid using direct quotes or copying words used in the articles. Often, a literature review, which lists and comments on the works done to date in a particular area of scientific investigation, is a component of a larger scientific paper. The review of the literature often proceeds in chronological order based on the publication date of the source and thus may differ from the scientific review paper, which is typically organized around concepts or other categories.

3. When you have finished writing your paper, write an abstract (approximately 100 words long) of your paper in which you summarize the major points in your review (see Chapter 5 for a discussion of how to write abstracts).

SAMPLE SCIENTIFIC REVIEW PAPER: SCIENCE FORMAT (CBE)[1]

Seasonal Variation in the Histochemical
and Biochemical Properties of Bat Muscle

Sheldon J. Cooper

Term Paper

For

Mammalian Physiology
Phys 501
Fall, 1992

I. INTRODUCTION 1
II. MAMMALIAN SKELETAL MUSCLE FIBER
         TYPES 4
III. HISTOCHEMICAL VARIATION IN BAT
         MUSCLE 5
     A. Pectoralis muscle 5
     B. Accessory flight muscles 6
     C. Gastrocnemius muscle 7
IV. BIOCHEMICAL VARIATION IN BAT MUSCLE 8
     A. Pectoralis muscle 8
     B. Gastrocnemius muscle 8
     C. Cardiac muscle 9
V. SUMMARY 9
VI. LITERATURE CITED 11

I. INTRODUCTION

Bats have been documented as having the highest maximal aerobic capacity in mammals. Flying bats have maximal oxygen consumption (VO_{2max}) of 2.5 to 3 times greater than running mammals of similar body mass (Thomas 1975). The metabolic rate of hibernating bats falls well below resting metabolic rates. In the little brown bat (<u>Myotis lucifugus</u>), the oxygen consumption rate (VO_2) is markedly reduced (99.3%) at 2°C relative to 40°C

(Hill and Smith 1984). A "shutdown" of the circulatory system accompanies these reduced metabolic rates in bats. During hibernation, heart rates in the red bat (*Lasiurus borealis*) have been reported at 10 to 16 beats/min^{-1}.

Similar heart rates have been documented in the little brown bat (24 to 32 beats min^{-1}), the big brown bat (*Eptesicus fuscus*; 42 to 62 beats min^{-1}), and the social bat (*Myotis sodalis*; 36 to 62 beats/min^{-1}) (Hill and Smith 1984). In contrast, heart rates during flight in the big brown bat have been recorded as high as 1097 beats/min^{-1} (Studier and Howell 1969). The muscle fiber composition of bat muscle and the metabolic system that provides ATP to the contractile system for fuel should reflect the extreme metabolic capabilities in bats (Armstrong et al. 1977; Brigham et al. 1990). Because bats undergo extreme physiologic changes during hibernation relative to summer, bats provide a good model for studying seasonal variation in muscle fiber adaptation and enzymatic adaptation to altered function. This paper examines changes in histochemical and biochemical properties of bat muscle during hibernation compared with summer.

II. MAMMALIAN SKELETAL MUSCLE FIBER TYPE

Skeletal muscle of mammals is composed of three distinct types of muscle fibers. In a given muscle these three types can be determined by their histologic staining properties. The different staining properties are based on differences in the chemistry of myosin ATPase in muscle fibers and on the amount of oxidative enzymes such as succinic dehydrogenase (Eckert et al. 1988). Slow-twitch oxidative fibers (SO) occur in postural muscles of mammals, such as the gastrocnemius muscle. SO fibers are slow contracting and slow fatiguing fibers that utilize aerobic oxidative metabolism. Fast-twitch oxidative fibers (FO) use oxidative metabolism as their primary energy source and are specialized for rapid repetitive movements. They are prominent in flight muscles of birds. Fast-twitch glycolytic fibers (FG) are powerful fibers that fatigue rapidly. FG fibers use glycolytic metabolism for their energy source (Foehring and Hermanson 1984; Eckert et al. 1988; Brigham et al. 1990). However, muscle fiber type is not a constant inherent characteristic. It appears that muscle fiber type depends on how the muscle is used (Schmidt-Nielsen 1990). Thus, it appears possible that bat muscle fiber composition may be capable of change depending upon use. Thus, bat muscle fiber composition may be different in summer relative to hibernation in winter.

III. HISTOCHEMICAL VARIATION IN BAT MUSCLE

A. Pectoralis Muscle

The pectoralis muscle is the primary flight muscle in bats. In summer, when bats are active, the pectoralis muscle and accessory flight muscles generate tremendous amounts of power daily. However, during hibernation these muscles are inactive except during brief periods of arousal (Armstrong et al. 1977). In spite of hibernation's effect on bat physiology and behavior, bats maintain the capability of flight during hibernation (Brigham et al. 1990). Thus, the question of seasonal variation in fiber composition in the pectoralis muscle needs to be examined.

Free-tailed bats (<u>Tadarida brasiliensis</u>) live in southeastern United States and undergo long-distance flights (Foehring and Hermanson 1984). Foehring and Hermanson (1984) reported that the pectoralis muscle of free-tailed bats is composed entirely of small FO fibers in summer. The little brown bat also shows 100% FO fibers in the pectoralis muscle in summer and also during hibernation (Armstrong et al. 1977; Brigham et al. 1990). Thus, it appears that in bats, diminished activity during hibernation does not change the fiber composition of the pectoralis muscle.

B. <u>Accessory Flight Muscles</u>

Foehring and Hermanson (1984) reported that free-tailed bats have SO fibers in both the triceps brachii complex and in the biceps brachii. The lateral head of the triceps was composed of 24% SO fibers and 76% FO fibers. The long head of the biceps was composed of 40% SO fibers and 60% FO fibers. Armstrong et al. (1977) found three distinctive FO fibers in the little brown bat's accessory flight muscles (the acromiotrapezius muscle and the acromiodeltoideus muscle). These include the following: FO_h fibers, having the darkest staining intensities for both ATPase and NADH-D; FO_m fibers, having an intermediate stain for the two enzymes; and FO_l fibers, having the lightest staining intensities. It has been proposed that the SO fibers in the triceps and biceps serve a tonic role for wing stabilization and that the FO fibers allow for rapid flexion or extension of the humerus, or possibly to provide force for humeral adduction during the wings downstroke (Foehring and Hermanson 1984).

C. <u>Gastrocnemius Muscle</u>

The muscle fiber composition of the gastrocnemius muscle in the little brown bat has been shown to be significantly different between active and hibernating bats. Gastrocnemius muscles of hibernating bats were composed of 62% FO fibers

(38% SO fibers) relative to 85% FO fibers (15% SO fibers) in active little brown bats. This fiber composition variation is possibly due to the lessened need for FO fibers during hibernation when bats are relatively inactive. Also, by hanging by the feet for long periods of time, bats are undergoing low-level chronic stimulation, which can shift fast fiber to slow fiber composition. However, since bats maintain the capability of flight, FO fibers are needed in the gastrocnemius muscle (Brigham et al. 1990).

IV. BIOCHEMICAL VARIATION IN BAT MUSCLE

A. <u>Pectoralis Muscle</u> The fiber composition of bat pectoralis muscle remains constant (FO) relative to season. However, does the glycolytic and oxidative potential of the pectoralis muscle change? According to Brigham et al. (1990), the glycolytic potential (measured as amount of phosphofructokinase) of the pectoralis muscle in the little brown bat was reduced 23.5% during hibernation. Also, the oxidative potential (measured as amount of citrate synthase) decreased by 15.2% during hibernation.

B. <u>Gastrocnemius Muscle</u> The fiber composition of bat gastrocnemius muscle has been shown to be significantly different during hibernation. The oxidative potential and glycolytic potential of gastrocnemius muscles in the little brown bat has been shown to decrease 56.5% and 60.5%, respectively, during hibernation relative to summer.

C. <u>Cardiac Muscle</u> It has been proposed that "oxidative potential of mammalian cardiac muscle should increase with increasing heart rate, while glycolytic potential should not (Ianuzzo et al. in press, cited in Brigham et al. 1990, p. 184). Brigham et al. (1990) reported that the little brown bat's glycolytic potential of cardiac muscle was constant regardless of season and that oxidative potential decreased 22.0% during hibernation. Thus, it appears that the hypothesis of Ianuzzo et al. (in press) holds true for the little brown bat.

V. SUMMARY

The structural and metabolic properties of bat muscle fibers demonstrate the extraordinary maximal metabolic rates capable in bats. However, many species of bats undergo hibernation during winter months, which drastically alters their physiologic processes. Muscle fiber composition and metabolic properties are somewhat plastic. Thus, during hibernation, both fiber composition and biochemical properties of muscles may change in bats. It has been shown that pectoralis muscle is homogeneous FO throughout the year in the little brown bat. It has also been found that both accessory flight muscles and gastrocnemius muscles have both FO and SO fibers in them. There is seasonal variation in the fiber composition of the gastrocnemius muscles of little brown bats. Biochemically, there is a general trend for reduced oxidative and glycolytic potential during hibernation in bats. These facts demonstrate that both muscle fiber composition and biochemical properties in bats can indeed change relative to season due to altered function.

VI. LITERATURE CITED

Armstrong, R. B.; Ianuzzo, C. D.; Kunz, T. H. 1977. Histochemical and biochemical properties of flight muscle fibers in the little brown bat, *Myotis lucifugus*. J. Comp. Physiol. 119:141-154.

Brigham, R. M.; Ianuzzo, C. D.; Hamilton, N.; Fenton, M. B. 1990. Histochemical and biochemical plasticity of muscle fibers in the little brown bat (*Myotis lucifugus*). J. Comp. Physiol. 160:B183-B186.

Eckert, R. D.; Randall, B.; Augustine, G. 1988. Animal physiology: Mechanisms and adaptations. 3rd ed. New York: W. H. Freeman and Company.

Foehring, R. C.; Hermanson, J. W. 1984. Morphology and histochemistry of flight muscles in free-tailed bats, *Tadarida brasiliensis*. J. Mamm. 65(3):388-394.

Hill, J. E.; Smith, J. D. 1984. Bats: A natural history. Austin, TX: Univ. of Texas Press.

Ianuzzo, C. D.; Blank, C. D.; Hamilton, S. N.; O'Brien, P.; Chen, V.; Brotherton, S.; Salerno, T. A. 1992. The relationship of myocardial chronotropism to the biochemical capacities of mammalian hearts. Biochemistry of Exercise VII. Champaign, IL: Human Kinetics Publishers.

Schmidt-Nielsen, K. 1990. Animal Physiology: Adaptation and environment. 4th ed. Cambridge, England: Cambridge Univ. Press.

Studier, E. H.; Howell, D. J. 1969. Heart rate of female big brown bats in flight. J. Mamm. 50:842-845.

Thomas, S. P. 1975. Metabolism during flight in two species of bats, *Phyllostomus hastatus* and *Pteropus gouldii*. J. Exp. Biol. 63:273-293.

NOTES

1. I am indebted to Professor James Gessaman of the Utah State University Biology Department for providing this student paper.

Index

Abbreviations
  of periodicals, 21
  using, 113
ABI-INFORM, 24
Abstract, of research paper, 123
Abstracts
  on CD-ROM, bibliographic references, 148
  computerized, focused search using, 133
  definition of, 12
  electronic, focused search using, 135
  focused search using, 133
  online, bibliographic references, 148
  for science and technology, 29–31
The ACS Style Guide: A Manual for Authors and Editors, 143
Active voice, versus passive voice, 109
Aerospace, indexes and abstracts for, 29
Agriculture, indexes and abstracts for, 29
Allied health, indexes and abstracts for, 30
Almanacs
  definition of, 12
  for science and technology, 27
Animal science, indexes and abstracts for, 29
Annotating, from sources, 49–54
Appendix, to research paper, 124
Applied Science and Technology Index, 134
Archimedes, 3
Article(s)
  magazine, bibliographic references, CBE style, 146
  newspaper
    locating, 20–21
    signed, bibliographic references, 146
    unsigned, bibliographic references, 146
  in periodicals, locating, 20–21
  in professional journals, locating, 21–23
Astronomy, indexes and abstracts for, 29
Atlases
  definition of, 12
  for science and technology, 27
Audience, 102–103
Author searches, of online catalogs, 16
Author/year documentation, CBE, for science and technology, 139–141

Batty, J. C., 99n
Bibliographic references. *See also* Documentation; Reference list; Sources; Working bibliography

abstracts
  on CD-ROM, 148
  online, 148
  annotated, 123–124
articles, 145–146. *See also* Bibliographic references, journal article; Bibliographic references, magazine article; Bibliographic references, newspaper article
  on discontinuous pages, 145
  with no identified author, 145
books, CBE style, 143–145
  all volumes in multivolume work, 145
  chapter in, 144
  with corporate author, 144
  with editor(s), 144
  one author, 143
  section in, 144
  selected pages in, 144
  two or more authors, 144
  two or more books by same author, 144
  work known by title, 145
dissertation or thesis, unpublished, CBE style, 147
electronic correspondence, CBE style, 148
e-mail, CBE style, 148
format for, 123
government document, 146
incorporation into paper, 117–118
internal citations of
  author/year system (CBE), 139–141
  multiple, 140
  with multiple authors, 140
  for one work cited in another work, 140
  with two authors, 140
  for two or more works by same author in same year, 141
  with no identified author, 145
  number system, 141
  multiple, 141
  one work cited in another work, 140
interviews, 147
journal article
  one author, 145
  online, 148
  two or more authors, 145
letters, 147
magazine article, CBE style, 146
movie, 147

newspaper article
 signed, 146
 unsigned, 146
paper presented at conference,
 unpublished, 147
reference work, 148
software for, 116–117
technical reports, 146–147
 corporate author, 146
 individual author, 146
Bibliographies, definition of, 12
Biography(ies)
 definition of, 12
 in science and technology, 28
Biological Abstracts, 135
Biology, indexes and abstracts for, 29
Book(s), bibliographic references. *See*
 Bibliographic references, books
Botany, indexes and abstracts for, 29
Brackets, 114
 in direct quotations, 118

Capitalization, 112–113
 and direct quotations, 118
CART. *See Current Contents Articles*
Cataloging systems, in libraries,
 computer, 13–19
CBE. *See Council of Biology Editors Style Manual*
CD-ROM
 abstracts on, bibliographic references, 148
 databases, searching, 24
 focused, 133, 135
Chapter(s), bibliographic references, CBE style, 144
Chemistry, indexes and abstracts for, 29
Citation indexes, focused search using, 134–135
CJOU. *See Current Contents Journals*
Cole, K. C., "The Scientific Aesthetic," 50–54
College research, 1–10
Colon
 to introduce quote, 119
 spacing after, 122
 using, 112
Comma
 to introduce quote, 119
 with quotation marks, 118
 using, 112
Computer(s). *See also* Word processing
 formatting and printing with, 119
 indexes and abstracts for, 29
Computer cataloging systems, in libraries, 13–19. *See also* Online catalogs
Computer research notebook, 35, 128, 137
Cooper, Sheldon J., 128

"Seasonal Variation in the
 Histochemical and Biochemical
 Properties of Bat Muscle,"
 150–161
Correct Grammar, 120
Council of Biology Editors Style Manual, 143
 model references, 143
Critical scientific research, 8–9
Critique, of sources, 56, 59–61
Current Contents, 131
Current Contents Articles [database], 14
Current Contents Journals [database], 14

Dart, Marion D., *Preliminary Design of
 Thermal Coupling Switch to Use
 Aboard Spirit II*, 82–98
Databases
 online services, 24, 25f
 searches
  CD-ROM, 24
  online, 24, 25f
Dictionaries, definition of, 12
Dissertation, unpublished, bibliographic
 references, CBE style, 147
Documentation. *See also* Bibliographic
 references
 format, software for, 116
 of reference materials, 116–117
 in science and technology, 138–143
  content notes, 141
  internal citation, 139–141
  author/year system (CBE), 139–141
  number system, 141
 reference list, 142–143
 style guides for, 143

EARS (acronym), 105
Easterbrook, Gregg, "The sincerest
 flattery. Thanks, but I'd rather
 you not plagiarize my work,"
 44–46
Ecology, indexes and abstracts for, 30
Editing, for grammar, punctuation, and
 spelling, 110–115
 with word processing, 115–116
Editorial Research Reports, 34
Einstein, Albert, 9
Electronic correspondence, bibliographic
 references, 148. *See also* E-mail
Ellipses
 in direct quotations, 118
 period with, 118
E-mail, bibliographic references, 148
Encyclopedias
 definition of, 12
 discipline-specific
  for science and technology, 27–28
  topic ideas from, 34

164 Index

Endnotes
  format for, 123
  software for, 116–117
  spacing, 121
Energy, indexes and abstracts for, 29
Engineering, indexes and abstracts for, 30
Engineering, research reports on, 81
  example of, 81–98
Entomology, indexes and abstracts for, 29
Environment, indexes and abstracts for, 30
Equipment evaluation reports, 81
  example of, 81–98
ERIC (Education Research Information Clearinghouse), 24
ERR. *See* Editorial Research Reports
et al., use of, 140
Exclamation point, with quotation marks, 119
Exercise(s)
  college research, 6
  library research methods, 38–39, 41, 49–54, 55, 61
  library resources, 31–32
  writing a review paper in science and technology, 129–130, 130–131, 132, 133, 135, 136, 137, 138, 149
Experiments
  laboratory, 63–64
  scientific, 63–64

Field observation and reports, *example of*, from field ornithology, 66–80
Field observations and reports, 64–66
Figures, using, 124–125
Film(s), bibliographic references, 147
First draft, writing, 138–139
Fisheries, indexes and abstracts for, 31
Fleming, Alexander, discovery of penicillin, 6–8
Food science, indexes and abstracts for, 30
Footnotes
  software for, 116–117
  typing, 123
Forestry, indexes and abstracts for, 30

General Book Collection [database], 13
General Periodicals Index [database], 13
General Science Index, 134
  format, 22, 22f
Geography, indexes and abstracts for, 30
Geology, indexes and abstracts for, 30
Geowriting: A Guide to Writing, Editing, and Printing in Earth Science, 143
Gessaman, James, 162*n*
Glossary of terms, 104
Gophers, 14
Government document(s)
  bibliographic references, 146
  locating, 23
Grammar, editing, 110–111
  with word processing, 115–116
Grammar-checking software, 116, 120
Grammatik, 116, 120
Graphics, using, 124–125

Handbooks
  definition of, 13
  for science and technology, 27
Hypotheses, formulation and testing of, 7–8

Illumination stage, of research, 3, 54–55
Illustrations, using, 124
Incubation stage, of research, 2–3, 33–54
Indexes
  computerized, focused search using, 133
  definition of, 13
  discipline-specific, 21, 26–31
    for science and technology, 29–31
  focused search using, 133
  to general-interest periodicals, 20
  of government documents, 23
  print, focused search using, 134–135
Index to Scientific Reviews, 131
Index to U.S. Government Periodicals, 23
Internet, 14–16
Interview(s), bibliographic references, 147
Italics, 113

Jargon, avoiding, 107
Journal(s), professional, locating articles in, 21–23
Journal articles, bibliographic references. *See* Bibliographic references, journal article
Journal articles, internal citations of, author/year system (CBE), 139

Keyword searching, of online catalogs, 18, 132–133
  advanced, 18–19, 19f

Lab experiments and reports, 63–64
Laboratory notebook, 64
Language, rhetorical decisions about, 103–104
LCSH. *See* Library of Congress Subject Headings
Letter(s), bibliographic references, 147
Library
  books, computerized catalogs, 13–19
  catalog
    computerized, 13–19. *See also* Online catalogs
    focused search of, 132–133
  other than your own, accessing information from, 26
  resources, 11–32, 27–55. *See also specific resource*

computer cataloging systems for, 13–19
online, 14–16
Library of Congress Subject Headings, 16
focused search using, 132–133
Library reference area, 11–20
search strategy for, developing, 129
Library research, 33–61
illumination stage, 3, 54–55
incubation stage, 2–3, 33–54
methods, 33–61
preparation stage, 2, 33–54
skills and principles, for review paper in science and technology, 127–128
verification stage, 3, 54–55
Library search
strategy for, 34
development of, 37–38, 128–130
time frame for, 39–41
Literature, guides to, for science and technology, 26–31

Magazine articles, bibliographic references, CBE style, 146
Magazine Index, 20
Margin(s), 121
justified, 120
Mathematics, indexes and abstracts for, 30
Medicine, indexes and abstracts for, 30
Medoc, indexes and abstracts for, 30
Microforms, 20
Mining, indexes and abstracts for, 30
Monthly Catalog of United States Government Publications, 23
Motion pictures. *See* Film(s); Movies
Movies, bibliographic references, 147

Newspaper articles
locating, 20–21
signed, bibliographic references, 146
unsigned, bibliographic references, 146
Newton, Isaac, 3, 9
Newtonian physics, 9
Nicolle, Charles, "The Mechanism of the Transmission of Typhus," 4–5, 10*n*
Nominalization(s)
changing, to action verbs, 108–109
definition of, 108–109
Notebook
laboratory, 64
research, 34–35, 128
on computer, 35, 128, 137
Notetaking, 43
from sources, 136
reviewing, 49
Numbers, writing, 114–115
Number system, for internal citations of sources, 141

Nursing, indexes and abstracts for, 30
Nutrition, indexes and abstracts for, 30

Observation(s), in sciences, 6–7
OCLC (Online Computer Library Center), 24–26
Online catalogs, 13
databases in, 13–14
main menu, 13–14, 15*f*
recording bibliographic information from, 19
searches/searching, 13–14, 38, 132–133
author, 16
keyword, 18
advanced, 18–19, 19*f*
subject, 16–17, 18*f*, 132–133
title, 16, 17*f*
subject headings, 16–19
Online database services, 24, 25*f*
Online networks, 14–16
Outline
of paper, constructing, 137–138
reverse, 106
Outlining, of sources, 49–54

Page numbering, in research paper, 122
Page references, citing, 57
PAIS. *See Public Affairs Information Services*
Paper presented at conference, unpublished, bibliographic references, 147
Paragraphs
focusing, 106–107
improving, 106–107
indentation for, 122
revising, 107
transitional sentence, 107
Paraphrases, documentation, 117, 138–139
author/year system (CBE), 140
Paraphrasing, 43–44, 57
acceptable versus unacceptable, 46–48
definition of, 43
Parentheses, 113–114
Penicillin, discovery of, 6–8
Period
with ellipses, 118
at end of sentence, spacing after, 122
with quotation marks, 118
Periodicals
abbreviations of, 21
citation information, copying, 21
definition of, 20
indexes to, 20
locating, 20–21
locating articles in, 20–21
Persona
definition of, 102
writer's, 102

Index

Pfeiffer, William S., 81
  Technical Writing: A Practical Approach, 99n
Physics
  indexes and abstracts for, 29
  Newtonian, 9
Plagiarism
  avoiding, 44–46
  definition of, 44
  examples of, 47–48
Preparation stage, in research, 2, 33–54
Primary research
  definition of, 63
  methods, 63–66
  in sciences, 63–66
    field observation and reports, 64–80
    lab experiments and reports, 63–64
    technology and engineering reports, 81–98
  time frame for, 39
Problem analysis reports, 81
  example of, 81–98
Problem-solving, stages of, 2–3
Professional journals
  citation information, copying, 21
  locating articles in, 21
Proofreading, 119–120
Public Affairs Information Services, 23
Punctuation
  with direct quotations, 118–119
  editing, 111–113
    with word processing, 115
Purpose statement, 102

Question marks, with quotation marks, 119
Quotation(s)
  direct
    documentation, 117, 138
    ellipses in, 118
    incorporation into paper, 117–119
    indentation, 118, 122
    introduction
      with colon, 119
      with comma, 119
    in quotation marks, 118
    use of, 117
  long, indentation for, 119, 122
Quotation marks
  for direct quotes, 57, 118
  exclamation point with, 118
  punctuation inside, 118
  question marks with, 118

Readers' Guide to Periodical Literature, 20
Reading
  active, 43
  for meaning, 41–43
Reference(s), library, definition of, 12
Reference area, beginning search in, 37–38

Reference list. *See also* Bibliographic references; Documentation
  format for, 123
  indentation for, 123
  in science and technology
    alphabetization, 142
    articles in, 142
    authors in, 142
    author/year system (CBE), 143
    books in, 142
    example of, 143–148, 159–161
    format for, 142–143
    journal names in, 142
    numbering, 142
    style guides for, 143
    titles in, 142
    two or more works by same author in same year in, 142
    volume and page numbers in, 142
    year of publication in, 142
  spacing, 121
  title for, 123
Reference work, bibliographic references, 148
Replicability, of scientific research findings, 9–10
Research
  college. *See* College research
  definition of, 1
  exercises. *See* Exercise(s)
  materials, 34–35
  methods
    library. *See* Library research
    primary. *See* Primary research
  process of, 2–3
    example: physician's use of, 4–5
    illumination stage, 3, 54–55
    incubation stage, 2–3, 33–54
    preparation stage, 2, 33–54
    verification stage, 3, 54–55
  in science and technology. *See* Science and technology; Scientific research
  in social sciences. *See* Social sciences
Research notebook, 34–35, 128
  on computer, 35, 128, 137
Research paper
  abstract of, 123
  appendix, 124
  format
    indentation, 122
    margins, 121
    spacing, 121
    formatting and printing with computer, 119
  graphics, 124–125
  incorporation of reference materials into, 117
  organization, 104
  page numbering, 122

planning, 101–104
proofreading, 119–120
revising, 104–120
title, 121–122
typed, 120
Research report
  on field observations, format, 65
  organization, 104
  in science and technology, 127–128
  in technology and engineering, 81
    example of problem analysis and
      equipment evaluation, 81–98
    format, 81
    writing, 65–66
Resources in Education, 23
Restatements, documentation, 117,
  138–139
Reverse outlining, 106
Review(s)
  definition of, 13
  scientific, focused search in, 130–135
Review paper, in science and technology,
  127–162
  discipline-specific sources for, 130
  example of, 150–161
  focused search and, 130–135
  general sources for, 130
  library research skills and principles
    for, 127–128
  organizing, 136–139
  preparation for, 128
  search strategy for, developing,
    128–130
  writing, 136–139
    arranging the materials, 137–138
    first draft, 138–139
Revision
  of rough draft, 104–120
  for structure and style, 105–107
  with word processing, 115–117
Rewriting, using word processing,
  115–117
Rhetoric, definition, 101
Rhetorical situation, 101–104
  components of, 101
RIE. *See Resources in Education*
Robotics, indexes and abstracts for, 29
Rough draft, revising, 104–120

Science, general, indexes and abstracts for,
  30
Science and technology
  discipline-specific sources for, 26–31,
    130
  documentation in, 138–143
  research report in, 127–128
  review paper in, 127–162
    example of, 150–161
    focused search and, 130–135
    general sources for, 130

library research skills and principles
  for, 127–128
organizing, 136–139
preparation for, 128
search strategy for, developing,
  128–130
writing, 136–139
  arranging the materials, 137–138
  first draft, 138–139
Science Citation Index, 134
Sciences
  formulation and testing hypotheses in,
    7–8
  observation in, 6–7
  primary research in, 63–66
    field observations and reports, 64–66
    lab experiments and reports, 63–64
    technology and engineering reports,
      81–98
    technology and engineering research
      reports, 66–80
Scientific debate, 9–10
Scientific experiments, 63–64
Scientific method, 7–8, 63–64
Scientific research
  critical, 8–9
  findings
    reliability of, 9
    replicability of, 9
    validity of, 9
  reporting, 9–10
Semicolon
  spacing after, 123
  using, 112
Sentence(s)
  improving, 107
  revising, 107
  spacing between, 122
Serials
  definition of, 13
  popular periodicals, locating articles in,
    20–21
Shupe, Janene, *Differences in Morning and
  Evening Bird Abundance,* 67–80
Social Sciences Index, 21
Software
  bibliography, 116
  for endnotes, 116
  for footnotes, 116
  grammar-checking, 116, 120
Sources. *See also* Bibliographic references;
    Working bibliography
  active reading of, 43
  author, 42
  discipline-specific, for science and
    technology, 26–31, 130
  evaluation of, 54–55, 135–136
  general, for science and technology,
    26–31, 130
  identification of, 136

locating, 41
notetaking from, 43
  reviewing, 49
organizational plan of, perceiving, 49
organization of, 42–43
paraphrasing. *See* Paraphrasing
publication, 42
reading for meaning, 42–43
section-by-section summaries of, making, 48–49
synthesis of relationship among, 56, 57–59
working with, 41–54
writing from, 55–61
  critiquing, 56, 59–61
  summarizing, 56–57
  synthesizing, 57–59, 81
Spell-checking, 116, 120
Spelling, editing, 110–111
  with word processing, 115–116
Statistics, indexes and abstracts for, 30
Style Manual: For Guidance in the Preparation of Papers for Journals Published by the American Institute of Physics, 143
Subject matter, 103–104
Subject searches, of online catalogs, 16–17, 18f, 132–133
Sullivan, Kim, 99n
Summaries from reference materials, documentation, 117
Summarizing, from sources, 48–49, 49–54, 56–57
Synthesis, of relationship among sources, 56, 57–59

Tables, using, 124–125
Taton, Rene, *Reason and Chance in Scientific Discovery*, 6, 8, 10n
Technical reports, bibliographic references, 146–147
Technology, research. *See* Science and technology
Technology, research reports on, 81
  example of, 81–98
Textiles, indexes and abstracts for, 30–31
Thesaurus, word processor's, 116
Thesis, unpublished, bibliographic references, 147
Time frame, for research project, outlining, 39–41
Title, of research paper, 121–122
Title page, 121–122
Title searches, of online catalogs, 17f
Tone, rhetorical decisions about, 104
Topic, ideas for, finding, 33–34

Underlining, 113
  in sources, 49–54
U.S. Government Reports, Announcements and Index (NTIS), 23

Verb(s)
  action, versus nominalizations, 108–109
  active voice versus passive voice, 109
  content, 108
  editing, 107–109
  empty, 108
  tenses, with direct quotations, 118
Verification stage, of research, 3, 54–55
Veterinary science, indexes and abstracts for, 31
Vocabulary, revising, 107
Voice, active versus passive, 109

WHUM. *See Wilson Guide to Art Index and Humanities Index*
Wildlife, indexes and abstracts for, 31
Wilson Guide to Applied Science and Technology Index, Biological and Agricultural Sciences Index, and General Sciences Index [database], 14
Wilson Guide to Art Index and Humanities Index [database], 14
Wilson Guide to Business Periodicals, Education Index, and the Social Sciences Index [database], 14
Word(s)
  commonly misused, 109–110
  improving, 107–110
Word processing
  editing with, 115–116
  formatting and printing with, 119
  revision with, 115
  rewriting using, 115–117
Working bibliography, 34, 35–37, 130
Works cited, format for, 123
Writer
  persona, 102
  purpose, 101–102
Writer's Helper, 120
Writing
  from sources, 55–61
    critiquing, 56, 59–61
    summarizing, 56–57
    synthesizing, 56, 57–59
  time frame for, 40–41
WSCI. *See Wilson Guide to Applied Science and Technology Index, Biological and Agricultural Sciences Index, and General Sciences Index*
WSOC. *See Wilson Guide to Business Periodicals, Education Index, and the Social Sciences Index*

Young, Richard E., et al., "The Four Stages of Inquiry," 10n

Zoology, indexes and abstracts for, 29